Experimental Design and Analysis for Tree Improvement

E.R. Williams, C.E. Harwood and A.C. Matheson

A catalogue record for this book is available from the National Library of Australia.

ISBN: 9781486317103 (pbk)
ISBN: 9781486317110 (epdf)
ISBN: 9781486317127 (epub)

Published in print in Australia and New Zealand, and in all other formats throughout the world, by CSIRO Publishing.

CSIRO Publishing
Private Bag 10
Clayton South VIC 3169
Australia

Telephone: +61 3 9545 8400
Email: publishing.sales@csiro.au
Website: www.publish.csiro.au
Sign up to our email alerts: publish.csiro.au/earlyalert

A catalogue record for this book is available from the British Library, London, UK.

Published in print only, throughout the world (except in Australia and New Zealand), by CABI.

ISBN 9781800624504

CABI
Nosworthy Way
Wallingford
Oxfordshire OX10 8DE
UK
Tel: +44 (0)1491 832111
Email: info@cabi.org
Website: www.cabi.org

CABI
We Work
One Lincoln Street, 24th Floor
Boston, MA 02111
USA
Tel: +1 (617)682-9015
E-mail: cabi-nao@cabi.org

Front cover: (top) Pineapples growing under a *Eucalyptus camaldulensis* plantation (photo by Khongsak Pinyopusarerk); (bottom, left to right) *Casuarina equisetifolia* trial at Ratchaburi, Thailand (photo by Khongsak Pinyopusarerk), *Casuarina junghuhniana* aged seven years growing near Bangkok, Thailand (photo by John Turnbull), Petri dishes of seeds in a germination cabinet (photo by Kron Aken)

Photos throughout are by the authors unless noted otherwise.

Cover design by Cath Pirret
Typeset by Envisage Information Technology
Printed in China by Leo Paper Products Ltd

CSIRO Publishing publishes and distributes scientific, technical and health science books, magazines and journals from Australia to a worldwide audience and conducts these activities autonomously from the research activities of the Commonwealth Scientific and Industrial Research Organisation (CSIRO). The views expressed in this publication are those of the author(s) and do not necessarily represent those of, and should not be attributed to, the publisher or CSIRO. The copyright owner shall not be liable for technical or other errors or omissions contained herein. The reader/user accepts all risks and responsibility for losses, damages, costs and other consequences resulting directly or indirectly from using this information.

CSIRO acknowledges the Traditional Owners of the lands that we live and work on and pays its respect to Elders past and present. CSIRO recognises that Aboriginal and Torres Strait Islander peoples in Australia and other Indigenous peoples around the world have made and will continue to make extraordinary contributions to all aspects of life including culture, economy and science. CSIRO is committed to reconciliation and demonstrating respect for Indigenous knowledge and science. The use of Western science in this publication should not be interpreted as diminishing the knowledge of plants, animals and environment from Indigenous ecological knowledge systems.

The paper this book is printed on is in accordance with the standards of the Forest Stewardship Council® and other controlled material. The FSC® promotes environmentally responsible, socially beneficial and economically viable management of the world's forests.

MIX
Paper | Supporting responsible forestry
FSC® C020056

Aug23_01

Contents

Acknowledgments

We are grateful to the Australian Centre for International Agricultural Research (ACIAR) for supporting the production of this book.

We thank Paul Ryan, Debbie Solomon, Robin Cromer, Vitoon Luangviriyasaeng, Maria Ajik, Khongsak Pinyopusarerk and Nguyen Duc Kien for allowing us to use their data sets; Matti Haapanen for the SAS programs; Muhammad Yaseen, Sami Ullah and Kent Eskridge for the R programs; Kron Aken for his support in the production of book material; and Roger Arnold for helpful comments on sections that were revised substantially for the third edition.

Preface to the third edition

As the title suggests, this book focusses on experimental design and analysis suitable for forestry field and glasshouse trials. Since the last edition there have been some major developments in the construction of spatial designs and partially replicated designs and these design types have been added to Chapter 7. In forestry there has also been a greater focus on clonal trials; an example is included in Chapter 8.

Since the last edition there have also been substantial developments in menus within software packages; this has to some extent obviated the need for the data pre-processing package DataPlus and reference to this package has been removed. On the other hand there is now considerable interest in the statistical package R for data analysis. The computer code for the Genstat and SAS analyses of the examples has been moved to online Appendices C and D respectively and a link to the analyses using R has been provided in the online Appendix E (see https://www.publish.csiro.au/book/8100#supplementary).

Whilst the terminology and examples are in the tree breeding context, the experimental designs and analyses can be applied to all areas of experimentation and in particular breeding programs for other plant species.

1

Introduction

1.1 Overview

This book is directed towards the preparation, design, analysis and interpretation of forestry trials, primarily for tree species/variety evaluation and genetic improvement. The structure of the book follows the sequence of thought needed by an experimenter to carry out a program of field trials for the purpose of comparing different genotypes (species, provenances, families, clones etc.). Relevant statistical methodology and computing techniques are introduced at appropriate stages throughout the book. The terminology of forestry field trials and examples from forestry are used so as to make the book particularly relevant to tree breeders. The statistical methodology and computing techniques, however, are applicable in other areas such as cereal breeding.

The book originated from a series of two-week courses given by Emlyn Williams in China (with Colin Matheson), Thailand and Kenya on behalf of the Australian Centre for International Agricultural Research (ACIAR). The experience gained from these courses has allowed the book to be structured in such a way as to be of most use to experimenters who are involved in forestry field trials. Since 1994, when the first edition was published, the book has formed the basis of many other courses given by Emlyn Williams, Chris Harwood and Kron Aken from the previous CSIRO Forestry and Forest Products, Canberra. Over the three decades since the first edition there have been major developments in tree improvement strategies. New genetic tools and increases in computing power and software are available to forestry researchers, but the requirements for well-designed field trials and their accurate analysis and interpretation remain.

We offer procedures for the various activities involved in design and analysis of field trials based on our experience with such trials in several countries over many years. Background theory on statistics and genetics is largely avoided, but references are given for further reading. Many relevant and recent examples are included, together with online Appendices and links to software programs for appropriate statistical analyses (see https://www.publish.csiro.au/book/8100#supplementary).

1.2 Outline

Chapter 2: Experimental planning and layout

In planning a field trial, considerations commonly include defining the experimental objectives, using appropriate sampling strategies to obtain seedlots or other experimental treatments such as clones, organising the experimental material and choosing the experimental design. These requirements are covered in this chapter. Simple experimental designs such as completely randomised and randomised complete block designs are introduced. Analyses of variance are carried out using the statistical package Genstat. In the book we will mainly use the term 'seedlots' when discussing forestry-related examples. The experimental details would, however, apply equally to 'clones' say, or in other contexts to treatments in general.

Chapter 3: Data collection and pre-processing

Efficient collection of data is very important in any field trial program. This is particularly the case in forestry experiments where it is usual to have multiple-tree plots and hence more items of data than in, say, a comparable cereal trial. Without an effective system for recording and pre-processing field data it is very easy for a data backlog to develop, often due to lack of training in handling such data rather than to the commonly stated reason, namely, staff shortage. Other aspects of data pre-processing involve simple data screening; this is very important to pick up transcription errors and to check statistical assumptions.

Chapter 4: Experimental design

This chapter describes several orthogonal designs using elemental model specification. The simple designs in Chapter 2 are treated in more detail, together with split-plot and factorial designs. The blocking and treatment structures in these designs are identified and lead directly to the analysis of variance.

Chapter 5: Analysis across sites

In evaluating treatments, experiments are usually conducted at a number of sites. There are several approaches to the analysis of data over sites. One could uncritically enter all data into a computer package and accept any answers that come out. However, it is better to break the process up into stages, as is proposed in Chapter 3, where individual tree data is summarised to a plot level before statistical analysis. One should also analyse experiments at each site separately and combine results in a two-way table of estimated treatment means by sites. Other information such as residual mean squares should be retained from the individual analyses for checking statistical assumptions. If required, the various stages of the analysis can then be combined in an overall analysis.

The two-way table of estimated treatment means by sites can be further analysed. If the site-by-treatment interaction is significant, there are many methods available to interpret this interaction, commonly known as genotype-by-environment interaction. One simple method is that of joint regression analysis, although the method becomes more complicated if the treatments used at each site are not the same. Methodology for handling this situation is introduced, together with programs for analysis. Other techniques for the analysis of genotype-by-environment interaction are briefly discussed.

Chapter 6: Variance components and genetics concepts

This chapter introduces the idea of mixed and random effects models, estimation of variance components and calculation of heritabilities. Other genetics concepts such as genetic correlations and response to selection are also discussed. Variance components are estimated using residual maximum likelihood (REML).

Chapters 1–6 allow the experimenter to follow the sequence of steps involved in the design and analysis of field trials, including combining results from trials conducted over several sites. However, for trials with more than about 16 seedlots in multiple-tree plots, it may be necessary to impose a greater degree of control over field variation within a trial than that provided by randomised complete block designs. This leads to the use of incomplete block designs and requires a more detailed statistical presentation. The final two chapters of the book describe the design and analysis of incomplete block designs.

Chapter 7: Incomplete block designs

The general theory of experimental design is introduced using a matrix approach. This level of complexity is needed to describe generalised lattice designs effectively. These designs are extremely useful in field trials with large numbers of treatments; their advantages over randomised complete block designs have been demonstrated repeatedly. The class of generalised lattice designs includes square and rectangular lattice designs and alpha designs. When plots are arranged in a rectangular grid, it is usually advantageous to use row–column designs with spatial enhancement. Partially replicated designs for initial screening trials are discussed. The computer package CycDesigN is introduced for the construction of all of these design types.

Chapter 8: Analysis of generalised lattice designs

Analysis of incomplete block designs is complicated by the fact that not all treatments appear in each block. This is best handled by using a mixed-effects model analysis and REML estimation. Specification of genetic treatments as random effects in the context of incomplete block designs, and strategies for effective comparison of controls with the main set of experimental treatments are demonstrated.

Appendix A: Introduction to Genstat

Genstat is a very powerful package which covers many different types of statistical analysis and has a large suite of commands. Because *Experimental Design and Analysis for Tree Improvement* concentrates on experimental design and analysis, and adopts a rigid structure for construction of data files, only a small subset of these commands is needed. These commands are discussed in this Appendix.

Online Appendices*

Appendix B contains several larger data files used in Chapters 5, 6 and 7. Appendices C and D list the Genstat and SAS code for the analysis of all the examples used in the book. The link to the R code for these examples can be obtained at Appendix E.

1.3   Software

1. The statistical package Genstat is used for analysis, including orthogonal and non-orthogonal analysis of variance, estimation of variance components and analysis of mixed-effects models. The statistical packages SAS and R can also be used for analysis purposes.
2. CycDesigN is used to construct and randomise a range of efficient incomplete block designs including alpha designs. Row–column designs for the two-way control of field variation and latinised designs for use with contiguous replicates can also be generated. The package can also produce partially replicated designs for use in initial screening trials. With all these design types spatial enhancement of experimental layouts is possible.

---

*   Access Appendices B–E at: https://www.publish.csiro.au/book/8100#supplementary

1.4 Summary

This book provides a set of procedures for the experimenter to follow when planning, designing and analysing a series of field trials where different treatments are compared. Appropriate software packages are discussed and relevant examples are given. The book is written in the forestry context but the statistical methodology can easily be applied to other areas such as cereal variety trials.

We see this book as appropriate for an undergraduate course for tree or plant breeders and as a reference book for experimenters involved in the design and analysis of field, nursery and glasshouse trials.

2

Experimental planning and layout

2.1   Introduction

Before a field trial is laid out, there are many activities to be undertaken and decisions to be made. If local and/or exotic tree species are being evaluated, seed collections are required following sampling strategies to represent species, provenances or populations. In many breeding programs, a large number of seedlots of a single species are assembled according to some predefined pattern to satisfy the genetic objectives of the trial. Adequate numbers of seedlings must be raised in the nursery. A suitable trial site needs to be chosen and the seedling resource has to be matched to the site resource. Note that, in general, for simplicity we use the term 'seedling'. In clonal trials, genetically identical 'seedlings' of each clone are produced by vegetative propagation and are properly termed 'ramets'.

The choice of layout of the seedlings at the site depends on the type of trial (e.g. species/provenance, provenance/progeny, clonal etc.) and the duration of the trial. Careful thought needs to be given to how many seedlings should be included in each experimental plot. Blocking methods need to be used to control field trend and appropriate randomisation must be employed. These and other issues concerning experimental planning and layout are covered in this chapter.

2.2   Experimental objectives

2.2.1   Experimental purpose

Experiments do not begin with a set of material to test. They begin long before, with the design of a research strategy. Tree improvement trials typically form part of a genetic improvement strategy which involves recurrent cycles of selection and breeding (Eldridge *et al.* 1993, chapter 20; White *et al.* 2007, chapter 11). In tree improvement trials, we compare different treatments (seedlots or silvicultural treatments) and we test the null hypothesis that there is no difference between the performance of the treatments. Usually, we wish to obtain further information from the trial, e.g. ranking the seedlots, selecting the best genetic entries for further use, estimating wood production or calculating parameters such as genetic variances and covariances. Additional purposes (such as selective thinning to create seed orchards) may also be incorporated into the experimental design as far as possible, but not to the extent of destroying one objective in the search for another or, worse, rendering the experiment inconclusive for all purposes.

We must be sure when designing experiments that they are capable of detecting the presence of any differences between seedlots. We must also ensure that these differences between entries are not due to other causes, such as environmental effects. Having determined that there are seedlot differences, we will probably wish to know which seedlots are best. 'Best' may refer to quantity (e.g. growth rate or fruit yield) or quality (e.g. stem form) for a variety of end uses. Most tree improvement trials test trees for timber, pulp or fuelwood production, but they may also test for pest or disease resistance, essential oil production or the use of foliage as animal

Agroforestry in Thailand: pineapples growing under a *Eucalyptus camaldulensis* plantation. (Photo Khongsak Pinyopusarerk)

fodder. Value for end uses is usually a combination of both quantity and quality traits, e.g. high value for pulp production is a combination of wood quantity as well as wood properties leading to high pulp yield and quality. These traits may be antagonistic, e.g. wood of high density is required for strong kraft pulp, but in many pine species the rate of growth in volume is negatively correlated with density, so the best seedlot for volume growth rate is unlikely to be best for high density.

2.2.2 Combining objectives

Traits involving survival present some difficulties for design. An experiment may be designed to test for survival, perhaps as resistance to frost or disease. To test adequately for differences, we must expect to kill about half the trees in the trial. This is because resistance to a stress is not linear and best discrimination between entries will be at about 50% overall mortality. It is not common to design trials to assess resistance to a stress in combination with other traits because this would lead to the loss of half the trees in the experiment, which would disturb other measurements. Survival is likely to be patchy, leaving spacing and therefore growing conditions uneven. However, in trials designed for other purposes survival is often used as a variate for analysis although the trial was not specifically

Alley cropping of tea with rubber trees, Hainan Is., China.

designed to test it. In some experiments this may lead to good discrimination between entries, but in others there may be no discrimination at all.

Competitive ability is a trait which is difficult to include in an experiment. We may wish to test for differences in ability of seedlots to compete with weeds. Are weeds likely to be a problem in the plantations that use the results from our experiment? If we keep an experiment completely free of weeds, the results may not apply to plantations which are less frequently weeded. As another example, results from clone trials using small plot sizes with considerable competition between adjacent clonal plots may not match well to clone rankings and per-hectare wood volume yields in monoclonal plantations (Stanger *et al.* 2011).

Coppicing ability may be an important trait for community forestry. Users may require fuelwood to be of small diameter so trees can be cut and handled easily. Village fuelwood lots may also be harvested at an earlier age than industrial plantations and then managed by coppicing on a short rotation. It is not possible to measure coppice growth at the same time as growth of the initial seedling crop; they must be assessed sequentially.

2.2.3 Trial duration

The length of time a trial is to run is important for design. Competition between adjacent plots is usually unimportant until the trees close canopy. Canopy closure

An agroforestry experiment in eastern Thailand involving dryland rice and *Acacia mangium*. (Photo John Turnbull)

may be delayed by planting trees further apart. Species elimination trials are often conducted using line plots in order to investigate the broad suitability of seedlots for a particular area. These trials are usually concluded before competition becomes a problem. Trials designed to last longer should have plots which feature square or rectangular arrays of trees. The effect of competition on the results can be reduced by using the outer trees in the plot as buffers and analysing only the inner trees, or net plot. In such cases plots should be at least 4 × 5 trees to ensure that net plots contain an adequate sample of trees.

2.3 Sampling strategies

Resources of labour, finances and available land are always limited. Choosing an appropriate set of seedlots to test is an important aspect of good experimental design. Thought must be given to the different kinds of material under test, whether they be clones, open- or control-pollinated families, provenances or species (Eldridge *et al.* 1993, chapter 3). The objectives of the experiment are likely to be different for different levels in the genetic hierarchy, and sampling strategies will also differ.

2.3.1 Natural populations

In genetics, a population is a group of individuals within which there is gene exchange. Although sometimes used interchangeably, population and provenance are not the same. Foresters usually define provenance as the geographic place of origin of a population of seed or plants, or as the population of plants growing at a particular geographic location (Turnbull and Griffin 1986). In other words, 'provenance' refers to the place of origin without the gene-flow implications of 'population'. It is assumed that natural provenances have each been subject to selection from their particular set of local environmental conditions and so will often differ in performance when grown at a common test site.

Within a species, we may think of a hierarchy of provenances grouped into regions and, in some cases, subspecies to make up the species. We may want to test one or a few provenances representing each of a large number of species in a species elimination trial. Alternatively, we may want to test many provenances of a particular species in a rangewide provenance trial to determine the best provenances of the species. Careful sampling, both to fairly represent each provenance tested and to choose the appropriate provenance or provenances for testing, is required. Guidelines for seed collection in tree improvement work have been comprehensively reviewed by Willan *et al.* (1990).

When sampling within a provenance it is important that we sample seed from a number of different trees. A sample of seed from just one tree, or a few closely related trees, may be unrepresentative, i.e. not a random sample of the trees that

This is an example of an inadequate base population of *Eucalyptus camaldulensis* at Xuan Khanh, Vietnam. On the left are trees from Petford, aged four years, outgrowing the trees on the right from a local source, aged five years. A change to growing trees from a better provenance will result in immediate improvement in productivity. (Photo John Turnbull)

make up the provenance. Australian researchers (Boland *et al.* 1980) generally recommend 10-tree bulk seedlots as sufficient to evaluate a provenance in trials, with the proviso that individual seed trees should be separated from one another by at least 100 m to minimise the extent of common descent, and that the bulk be made up of approximately equal numbers of viable seeds from each seed tree. A standard of 25 trees per provenance was specified for the FAO/IBPGR (Food and Agriculture Organization/International Board for Plant Genetic Resources) collections of arid-zone species (Palmberg 1981), but it may be hard to collect 25 unrelated trees per provenance for species with small local population sizes. Keeping the identity of seed from individual trees separate for use in provenance/ progeny trials is to be encouraged; samples from the individual seedlots can be combined later to provide a provenance bulk if required.

If we sample open-pollinated seed from seed trees in natural populations, we commonly assume that the samples (families) form half-sib groups. This is usually not true because there may be some self-pollination (Brown *et al.* 1975) and neighbourhood inbreeding (i.e. mating between nearby trees that are close relatives, Eldridge *et al.* 1993, chapter 19) and some unrelated trees may be represented as fathers more than once in a family (Squillace 1974). Open-pollinated families of pines may be closer to half-sibs than eucalypts or acacias (Matheson *et al.* 1989; Moran *et al.* 1989).

In selecting a single provenance seedlot of a well-known species for a species elimination trial, we might choose a provenance which has performed well in other, similar trial environments. An example would be the Petford provenance of *Eucalyptus camaldulensis* for a seasonally dry tropical environments, or the Lake Albacutya provenance of the same species for a site with a Mediterranean (warm temperate winter rainfall) climate. If the species is poorly known, we would commonly sample a natural provenance where it shows its best development, or, alternatively, a provenance from the region most closely matched to the test site. Computer-based climatic matching programs are now available for many countries to assist such decisions (Booth and Jones 1998).

For many species with extensive, more or less continuous, distributions, hundreds or thousands of natural provenances could be defined, so there needs to be a rational sampling approach to obtain a set of 10–30 or so provenances which can feasibly be tested in field trials. For the purpose of international provenance trials, a total of 51 provenances of *Calliandra calothyrsus* were collected from eight countries covering the entire geographic range and as much of the edaphic, biotic and climatic range of the species as possible (Macqueen 1993). A set of some 20 provenances, for which good quantities of seed were available, was tested at many international trial sites, with additional provenances selected for testing at particular trial environments or for particular end uses (fuel, fodder etc.).

Various strategies may be followed when assembling a collection of seedlots for a provenance trial. Green (1971) sampled *Eucalyptus obliqua* by superimposing a 70 mile (113 km) square grid on a map showing the known natural range of the species. A total of 22 general locations closest to the points of intersection of the grid

These *Eucalyptus globulus* seedlings are destined to become part of a provenance/progeny trial near Kunming, China. The trial is to be used as a seed source later. Uneven germination among seedlots meant that some seedlots had to be omitted from the trial despite all the effort that went into collecting the seeds.

lines was selected, and provenance collections made from the nearest undisturbed stands with good seed crops. This study was very successful at detecting patterns of variation in the species. Another successful study on *Eucalyptus delegatensis* by Moran *et al.* (1990) adjusted sampling density according to the natural abundance of the species. Places where *E. delegatensis* was most abundant were sampled more intensively.

Natural patterns of genetic variation in many species have become blurred through human interference. The geographic, edaphic and climatic range were surveyed and sampled in a representative way for provenance collections of *Gliricidia sepium* in Central America (Stewart *et al.* 1996). Provenance variation as a response to environmental variation appeared to have been heavily overlain by the effects of centuries of human activity, including possible movement of seed between provenances and local selection of desirable phenotypes. A total of 28 provenances from across the natural range of the species were collected for international trials.

2.3.2 Planted populations

The natural provenance which was the original source of seed for a planted stand is important in determining its performance and the genetic characteristics of the seed it produces. Two nearby planted stands will produce seed of differing genetic quality if they are derived from contrasting natural provenances, so it would be misleading to describe them as the same artificial or derived provenance.

The genetic history of a planted stands following the initial introduction is also important. Inbreeding, which occurs frequently in artificial populations because of a narrow genetic base or absence of suitable pollinating agents, leads to reduced vigour and poor form in many forest tree species (Sedgley and Griffin 1989). Seed collections from inbred stands, or from single trees in arboreta or botanical gardens, will perform poorly and will not fairly represent the true ability of the species. On the other hand, seedlots collected from plus-trees selected in plantations which have been established from a broad genetic base of superior provenances often perform better than the natural provenances because neighbourhood inbreeding has been eliminated and individuals well-adapted to the exotic environment are selected. Such collections in exotic plantations have formed the initial breeding populations of the *Pinus radiata* breeding programs in Australia and New Zealand. Frequently, tree breeders compare such selections in field trials together with commercially available seed sources and seedlots from natural provenances of the species (e.g. Pinyopusarerk *et al.* 1996).

The genetic relationships between and within seedlots produced by controlled pollinations in well-developed breeding programs are usually known, except for polycrosses in which pollen mixtures are used (Zobel and Talbert 1984).

Three-year-old eucalypt clone trial, Paraguay, showing clone performance in large blocks. On the left is a clone of *Eucalyptus grandis*; on the right a clone of the *E. grandis* × *E. camaldulensis* interspecific hybrid which has failed badly, displaying stem breakage and heavy insect attack.

With the increasing trend towards clonal forestry, it has become necessary to test large numbers of candidate clones to identify those giving the best performance in plantations. Other types of clone trials aim to rank individuals more accurately in the breeding population (Shelbourne 1992), or provide the accurate phenotyping required for molecular genetic studies. Although there is no genetic variation between the ramets which make up a clone (except for somaclonal variation in some tissue-cultured clones), non-genetic clonal effects ('C' effects) which result in differing physiological status and hence differing performance among ramets must be taken into account (Lerner 1958).

2.4 Allocation of resources

2.4.1 Seedling resource

In the previous section we discussed some methods for selecting seedlots. The next stage might be to carry out an experiment to compare seedlots. To do this the seeds must first be germinated in the nursery. The experimenter needs to know roughly how many seedlings of each seedlot will be required for a comparative trial. Then, using information on seed viability, we can ensure that enough seed is sown in the nursery. For example, if 80 seedlings are required for a seedlot and the viability as provided by the Australian Tree Seed Centre (ATSC) is 300 seedlings per 10 g, we would want to be supplied with at least 5 g of seed to allow for inevitable nursery losses. This also permits selection of healthy and even-sized germinants for experimentation and allows a few extra seedlings for refills in the first weeks of the trial. Note that we use the term 'seedlings' to refer to young trees regardless of whether they are derived from seed or clonal material.

Nursery conditions can vary considerably and, strictly speaking, we should look towards arranging the young plants in a designed experiment which can be analysed in its own right. Normally it is assumed that if there is any nursery variation it will not influence the results of the main field trial. Sometimes, however, there is large variation in seedling growth within seedlots. Seedlings should then be grouped into size classes, which are assigned to different replicates of the field trial. In this case we say that the different seedling size classes have been confounded with replicates.

The number of seedlots finally available for experimentation cannot be determined until after nursery germination. For example, we might start with 200 seedlots but under the nursery conditions only 180 provide enough seedlings to be included in a field trial. The seedlots should be labelled from 1 to 200 in the nursery and be linked via a key to the actual seedlot identities. It is important to

Casuarina junghuhniana aged seven years growing near Bangkok, Thailand. The need for deep drainage ditches shown here must be taken into account in trial design. (Photo John Turnbull)

maintain the nursery labels in the field trial even though we are only using 180 of them. This is especially so in multi-location trials where different nurseries attempt to raise the same set of seedlots. Using a common and unchanged system of field numbers, linked by the one key, greatly facilitates analysis across the different sites. The software package CycDesigN (see Chapter 7) is used to produce randomised field layouts and has a facility that enables the labels to be specified.

The seedlots to be tested might consist of a number of provenances and families within provenances, or different species and provenances within species. In such cases we say that there is a nested treatment structure in the seedlots (see Section 8.4). This structure should be incorporated in the field layout, and once again this can be done with CycDesigN. At this stage, however, we will consider the seedlots as though they had no structure, i.e. in the above example of 200 arbitrary seedlots no two seedlots have a closer genetic relationship than any other two.

Once the number of seedlots and seedlings per seedlot has been determined in the nursery, the question arises 'How should the seedlings of each seedlot be organised in the field trial?' We look at this question in the next section.

2.4.2   Site resource

The site for a field trial must be selected carefully. It is not essential for the site to be completely uniform, as the experimental design will help to remove systematic trends. However, we do not want a site that contains a lot of local variation (e.g. mounds, holes, rocks, ash-heaps etc.) as the statistical model adopted for the analysis of the field trial may not effectively handle such a situation, although a spatial design (Section 7.6) and analysis can help. For forestry trials it is common to seek a rectangular area on which a grid of planting positions can be superimposed. The spacing between trees at planting will vary depending on the nature of the field trial. For long-term trials a spacing of 3×3 m is often used, although subsequent thinning may be required to enable satisfactory stand development, particularly for sawlog production. Trials designed to run for only 3–5 years can be planted at 2×2 m spacing if above- or below-ground competition is not expected within that time. Long-term trials in dry climates might use a wider spacing of 4×4 m.

Elimination trials involving large numbers of seedlots may run for only two years and line plots at 1.5 m spacing with 3 m between rows are often used. This rectangular planting grid is also used when early thinning is planned; a line plot of five trees might be thinned down to three trees after two years and ultimately down to one tree if a progeny trial is converted into a seedling seed orchard. Sometimes site preparation includes mounding in rows 3 m, 4 m or 5 m apart; in such cases the spacing between trees on a mound is likely to be less than between rows.

Ultimately the site resource becomes an area of land with a grid of planting positions, large enough to accommodate the seedlings prepared for the field trial and, preferably, one or two buffer rows around the outside of the trial.

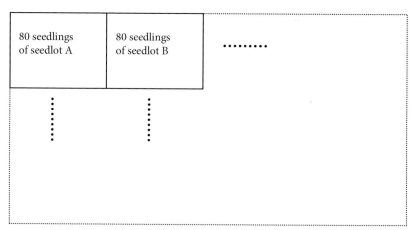

a) one replicate, internal replication of 80

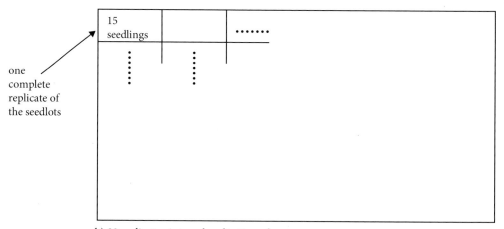

b) 80 replicates, internal replication of one

Fig. 2.1. Options for the allocation of 80 seedlings from 15 seedlots to planting positions in two experimental designs.

Let us suppose that we have 80 seedlings from each of 15 seedlots to be tested in a field trial. We have a total of $80 \times 15 = 1200$ seedlings to be allocated to the grid of planting positions. There are many ways that this can be done. Some possibilities are as follows:

1. Assign the 1200 seedlings completely at random to the 1200 planting positions. This is known as a completely randomised design. The problem with this arrangement is that we have no simple mechanism for making adjustments in the analysis to cater for any trends across the field site. It is very rare that a site is completely uniform. The most common reason for field variation or trend is a slope or gradient, e.g. trees may tend to grow better at

the bottom of a slope than at the top. But there are many other reasons for field variation, such as soil changes, prevailing wind, aspect effects and site preparation techniques. We need an experimental layout that allows us to account for field variation adequately, thereby leading to an accurate and efficient comparison of seedlot means. In Chapter 4 we will look at linear models for data, make assumptions about the nature of field variation and test these assumptions. The art of experimental design is to cater as well as possible for any field variation; a completely randomised design has no facilities for this purpose.

2. Another possibility would be to group all seedlings of each seedlot together, as illustrated in Fig. 2.1(a). With this arrangement there is internal replication of 80 (i.e. there are 80 seedlings in each plot) but no external replication. Chapter 4 shows that it is the amount of external replication that is important in allowing us to obtain an error term in the analysis of variance. Without external replication we cannot calculate an appropriate error term, so that the Fig. 2.1(a) arrangement is unsuitable for the statistical comparison of seedlot means.

3. At the other extreme is the arrangement in Fig. 2.1(b) where there is external replication of 80, each with just a single seedling from each seedlot (i.e. single-tree plots). A group of plots consisting of a single representative of each of the seedlots is called a replicate. In this book we will deal mainly with experimental designs in which external replication is via discrete replicates; these are called resolvable designs. Figure 2.1(b) clearly maximises the number of replicates and so apparently maximises the degrees of freedom available for the error term for testing seedlots (Chapter 4). If seedling survival is less than 100%, however, this single-tree plot arrangement can cause considerable difficulty with analysis. A missing tree here is a missing plot, thereby leading to a non-orthogonal analysis of variance which for a large number of replicates and seedlots can be quite demanding and there is the likelihood of reduced precision on estimated effects. Another disadvantage of single-tree plot arrangements is that there is no possibility of obtaining an estimate of the between-tree-within-plot variance component (Section 6.2.1). This quantity provides a good indication of genetic variation within seedlots, which is very useful in formulating genetic improvement strategies.

Although White *et al.* (2007, chapter 14) advocate single-tree plot designs for clonal and progeny tests on the grounds of statistical efficiency, survival rates below 95% can cause estimation problems with single-tree plots. In tropical regions overall survival of *Acacia* and *Eucalyptus* species may fall to as low as 70–80% by age 3–4 years when field trials are typically assessed (Kien *et al.* 2010; Son *et al.* 2018). We therefore recommend two-tree plots as a minimum plot size; this greatly reduces the number of missing plots at assessment. Apart from the statistical issues, having two or more trees per plot assists with correct data recording, providing a check on the recorder's exact location in trials where labelling of all individual plots is not feasible. The is particularly so for clone trials where individual clones are often distinguishable based on their leaf, stem and bark morphology. A comparison of single- and multiple-tree plots was conducted by Stanger *et al.* (2011).

4. The situation in Figs 2.1(a) and 2.1(b) represent extreme alternative ways of allocating the 80 seedlings of each seedlot to planting positions. In between there are other possibilities, e.g. 20 replicates of four-tree plots (internal replications), five replicates of 16-tree plots and many other combinations. The choice depends on several considerations. For a species/provenance trial it is better to have more trees per plot to help minimise effects of possible competition between adjacent plots of different species; with 80 seedlings per seedlot we would opt for either five replicates of 16-tree plots or four replicates of 20-tree plots. Fewer than four replicates is generally not advisable, unless seedlings or land are in very short supply. The trees in each plot would be arranged in a 4 × 4 or 4 × 5 grid; then if after several years there is a concern about above- or below-ground competition, an internal plot of 2 × 2 or 2 × 3 trees can be used for analysis and the results compared with those from the whole plot to quantify competition. Species/provenance trials typically run for at least five years and the seedlots should have been selected with prior knowledge that they are suitable for the trial site. Thus there is little prospect that a relatively large plot of 16 or more trees will fail completely, leaving a patchy trial.

Often, elimination trials are run for a short time to gain information on the suitability and performance of seedlots prior to more detailed longer-term trials. Elimination trials require only 20–40 seedlings per seedlot and use line plots of four to eight trees. As the number of seedlots in an elimination trial may exceed 100, a replicate can become quite large and there is the need to introduce extra blocking structures (such as incomplete blocks) within replicates. These are discussed in Chapter 7.

For species/provenance trials, longer-term provenance trials and elimination trials, the recommended number of replicates is four to six, although this depends on the number of available seedlings. For example, with 20 seedlings per seedlot available for an elimination trial, we would probably opt for four replicates of five-tree line plots. But if we had the luxury of 80 seedlings per seedlot for an elimination trial of 300 seedlots, it would be better to have either eight replicates of 10-tree plots or 10 replicates of eight-tree plots. The final choice often depends on the size and shape of the trial site; one of the acceptable configurations of replicates and plot size might fit naturally onto the trial site.

Another common type of trial is the progeny or provenance/progeny trial which is selectively thinned to a seedling seed orchard. The requirements for this type of trial are a compromise between what is desirable for the comparison of seedlots and the production of superior seed. From the statistical point of view, we want to have enough replicates (4–6) to provide adequate degrees of freedom for the error term used for testing, while at the same time ensuring that there are enough trees in each plot to protect against missing plots and to reduce, to some extent, competition from adjacent plots. A sensible approach is to select the orchard area and final working density of retained trees, and check that the initial spacing, design layout and selected thinning regime can generate the desired final configuration and accurately test the seedlots.

2.5 Strata

In the last section we looked at various ways that the seedlings of each seedlot can be allocated to planting positions. The completely randomised design does not impose any restrictions on this allocation; the seedlings can be randomly assigned to any position. On the other hand, arrangements where there is some internal and external replication require restrictions on the assignment of seedlings to planting positions. For example, for an experimental layout with five seedlots in four replicates of four-tree line plots (Fig. 2.2) we must ensure that the first four planting positions all have the same seedlot, then that the first five plots contain different seedlots to make up the first replicate, and so on. Each restriction imposed on the layout of the seedlings is called a stratum in the experimental design. In Fig. 2.2, one stratum would result from grouping the 80 planting positions into 20 plots of four seedlings. Another stratum results from grouping the 20 plots into four replicates each with five plots. In addition, there is a stratum at the tree or planting position level, so overall in Fig. 2.2 we have three strata – the replicate (*repl*) stratum, the plot-within-replicate (*repl.plot*) stratum and finally the tree-within-plot-within-replicate (*repl.plot.tree*) stratum.

As mentioned above, the reason for concerning ourselves with the very important concept of strata in experimental design is to accommodate and adjust for field variation. If this is done successfully, seedlot means are estimated and comparisons made more precisely. In Chapter 4 we will look at linear models that make assumptions about the nature of field variation. By defining strata we hope to model field variation more accurately and thereby account for its effect. Suppose that in Fig. 2.2 there is some field trend (e.g. a slope) from left to right. Without any strata we would have to assume that the variation between two planting positions on different sides of the site would be no greater than between two adjacent positions, clearly an unrealistic assumption. With the *repl* stratum, however, we have the facility to make adjustments between replicates to help accommodate the field trend from left to right so it does not distort the test of significance for seedlots. The ability to separate seedling variation from field variation by the use of models and assumptions (Chapter 4) is an integral part of the concept of strata in experimental design.

The strata we have defined for the Fig. 2.2 arrangement are nested; the trees are within plots which in turn are within replicates. The type of randomisation needed for this layout has to be consistent with this nested structure. The 16 seedlings of the five seedlots are allocated to the planting positions according to the following restricted randomisation:

Progeny trial of *Acacia auriculiformis* at 20 months of age growing at Chanthaburi, Thailand. It is planned to thin the trial for use as a seedling seed orchard after data collection is complete. (Photo Khongsak Pinyopusarerk)

1. The five seedlots are assigned at random to the five plots in replicate 1, then as a separate operation the seedlots are randomly assigned to the plots of replicate 2, and so on.

2. A random selection of four seedlings of each seedlot is randomly assigned to the four planting positions in each plot. There is sometimes a variation on this procedure when there are differences in size (or quality) of the seedlings: the seedlings of each seedlot might be grouped into four classes according to size (quality) and classes randomly assigned to the four replicates. The field variation between replicates would then be confounded with the variation between seedling classes. Here, rather than using experimental design and assumptions to separate field and seedling variation, we have deliberately mixed them up. Choosing when to confound or keep separate sources of variation is part of the art of experimental design.

Nested strata as in Fig. 2.2 are used to accommodate field variation in one direction. When variation is in two directions, crossed strata are used. In Fig. 2.3 we have a *row* stratum and a *column* stratum to cater for row and column variation. A third stratum is then the interaction between rows and columns

Replicate

Plot	1	2	3	4
1	x x x x	x x x x	x x x x	x x x x
2	x x x x	x x x x	x x x x	x x x x
3	x x x x	x x x x	x x x x	x x x x
4	x x x x	x x x x	x x x x	x x x x
5	x x x x	x x x x	x x x x	x x x x

direction of trend

Fig. 2.2. Layout of seedlings in four replicates of five plots, each with four trees (each x refers to one tree).

Column

Row	1	2	3	4	
1	x	x	x	x	
2	x	x	x	x	direction
3	x	x	x	x	of trend
4	x	x	x	x	

direction of trend

Fig. 2.3. Two-dimensional layout with four rows and four columns (each x refers to a whole plot).

(*row.column* stratum) corresponding to the plots in the layout. Finally, if each plot contains a number of planting positions, we would have the *row.column.tree* stratum.

The understanding of strata is fundamental to the use of the software package Genstat, and indeed to the understanding of experimental design itself. This book will make use of strata throughout. Notation for specifying strata in Genstat will be discussed in the next section, leading to specifications for analysis of variance tables.

2.6 Analysis of variance

In this section we will look at the analysis of variance of some simple designs. At this stage we will merely relate the concept of strata, as introduced in the previous section, to (i) the Genstat specification for blocking structures and (ii) the resulting analysis of variance table. In Chapter 4 we will consider experimental designs in more detail and demonstrate the importance of the underlying linear model in determining the appropriate analysis. Genstat syntax will be used in this section; an introduction to Genstat is given in Appendix A.

2.6.1 Completely randomised design

This design is discussed in Section 2.4.2; it amounts to assigning the seedlings completely at random to planting positions. As such there is only one stratum, the *tree* stratum. So the Genstat specification for the blocking structure (Appendix A) is:

block   tree

The seedlings provide external replication for the different seedlots. The Genstat treatment structure is:

treatment   seedlot

It is very rare for completely randomised designs to be used. Just the concept of multiple-tree plots immediately introduces at least two strata, the *plot* stratum and the *plot.tree* stratum. The completely randomised analysis can be encountered, however, when analysis of variance is carried out on plot means rather than individual tree data; this approach is discussed at length in the next chapter. If the external replications of the seedlots have been assigned at random to the plots, we would have the following Genstat specification for the completely randomised analysis of plot means:

block   plot

treatment   seedlot

Some experimenters have introduced the idea of non-contiguous plots (Libby and Cockerham 1980) where groups of trees planted apart are analysed as though they form a discrete plot. This would involve introducing into the analysis a *plot.tree* stratum which does not in fact exist and the method is thus not recommended.

Example 2.1

A field trial was planted to compare a seedlot derived from a seed orchard (SO) with one collected from a routine plantation (P). There were eight plots of each seedlot, thinned at seven years of age. Tree diameters at breast height (*dbh*) were measured at 15 years and plot means calculated (Table 2.1). A Genstat program to carry out a completely randomised analysis of plot means is given in Table C2.1; see Appendix A for further details about syntax.

Table 2.1 *Dbh* (cm) means for a seed orchard (SO) and a routine plantation (P) seedlot.

| | Seedlot | |
Replication	SO	P
1	30.38	28.16
2	27.91	25.62
3	28.06	26.61
4	31.42	32.59
5	30.11	28.80
6	31.52	28.19
7	31.72	31.23
8	33.53	28.34

Table 2.2 Genstat output from analysis of variance of *dbh* means for Example 2.1.

Analysis of variance

Variate: dbh

Source of variation	d.f.	s.s.	m.s.	v.r.	F pr.
plot stratum					
seedlot	1	14.270	14.270	3.25	0.093
Residual	14	61.410	4.386		
Total	15	75.679			

Tables of means

Variate: dbh

Grand mean 29.64

seedlot	SO	P
	30.58	28.69

The Genstat output is given in Table 2.2. Included in the output is the analysis of variance table and the estimated means for SO and P. We use the analysis of variance table to test the significance of differences between seedlots, i.e. whether the seed orchard seedlot is significantly different from the plantation seedlot. The test of significance is carried out by taking the ratio of the seedlot mean square (14.27) and the residual mean square (4.386). The variance ratio (3.25) is then compared with F-distribution tables on 1 and 14 degrees of freedom (d.f.), or we can simply get the F-probability printed as an option in the analysis of variance table. Here the F-probability value is 0.093, which is greater than the nominal 5% (0.05) significance level and so we would conclude that the seedlots are not significantly different.

2.6.2   Randomised complete block (RCB) design

When the external replication of seedlots is grouped into replicates, we have a randomised complete block structure. With multiple-tree plots there would be three strata, *repl*, *repl.plot* and *repl.plot.tree*. The Genstat specifications for block and treatment structures (Appendix A) are:

block   repl / plot / tree

treatment   seedlot

Single-tree plots, or analysis of plot means, would result in just two strata and the Genstat specification for the block structure would be:

block   repl / plot

Example 2.2

The plots in Example 2.1 were in fact laid out as an RCB design (Fig. 2.4) and so the Genstat program in Table C2.1 needs to be amended (Table C2.2). The resulting Genstat output is in Table 2.3. The important difference between the analysis of variance there and that given in Table 2.2, is that there is now an extra stratum to cater for variation between replicates. The replicate mean square is 6.981 compared with a new residual mean square of 1.792. The residual sum of squares in Table 2.3 (61.410) has been split into two components, the replicate sum of squares (48.867) and the new residual (12.543).

It is clear that the presence of *repl* as an extra blocking factor has accounted for much of the field variation and as a result the residual mean square has been substantially reduced. More importantly, the variance ratio (7.96) on 1 and 7 d.f. is now significant at the 5% level.

Examples 2.1 and 2.2 highlight two fundamental issues that contribute to the major thrust of this book:

				Replicate			
1	2	3	4	5	6	7	8
SO	SO	P	SO	P	P	SO	P
30.38	27.91	26.61	31.42	28.80	28.19	31.72	28.34
P	P	SO	P	SO	SO	P	SO
28.16	25.62	28.06	32.59	30.11	31.52	31.23	33.53

Fig. 2.4. Layout of plots for Example 2.2 with seedlot labels (SO or P) and *dbh* means.

Table 2.3 Genstat output from analysis of variance of *dbh* means for Example 2.2.

Analysis of variance

Variate: dbh

Source of variation	d.f.	s.s.	m.s.	v.r.	F pr.
repl stratum	7	48.867	6.981	3.90	
repl.plot stratum					
seedlot	1	14.270	14.270	7.96	0.026
Residual	7	12.543	1.792		
Total	15	75.679			

Tables of means

Variate: dbh

Grand mean 29.64

seedlot	SO	P
	30.58	28.69

1. The use of appropriate blocking in experimental layout can considerably reduce the residual mean square, thereby leading to a more precise analysis. By incorporating replicates into Example 2.2, the residual mean square was reduced from 4.386 to 1.792. So, even though the estimate of seedlot means did not change from the completely randomised to the randomised block design, the conclusions did change – a direct result of the reduced residual mean square. In Table 2.2 the analysis was not sensitive enough to demonstrate seedlot differences, whereas in Table 2.3 appropriate blocking via the *repl* stratum has resulted in statistically significant seedlot differences. In Chapters 4 and 7 we will look at many other types of experimental designs and blocking structures. In every case the objective is to control systematic variation, such as field trend, so that the residual mean square is as small as possible.

2. The data as presented in Table 2.1 give no indication whatever of the type of blocking structure appropriate for the analysis. Hence, we could very easily opt for the completely randomised analysis as done in Table 2.2. If, however, the data had been presented in conjunction with a field layout, as in Fig. 2.4, it would be clear that replicates and restricted randomisation had been used. Hence the RCB analysis in Table 2.3 is appropriate.

The next chapter will emphasise the value of knowing the field layout of a trial and recording data in an order reflecting that layout.

3

Data collection and pre-processing

3.1 Introduction

In Chapter 2 the design, layout and analysis of some simple experiments was discussed. This allowed the use of Genstat for analysis of experimental designs to be demonstrated. In practice, however, a lot of preparation is needed before data files are in the form where we can use Genstat to produce the analysis of variance table. To expand on what should be considered in the collection and preliminary processing of data, we will take a simple situation of an experiment with 49 seedlots and two replicates, giving a total of 98 plots, and suppose that in each plot there are 16 trees to be measured.

1. What is the best way to record height and diameter information for each tree?
2. Should data sheets or an electronic data capture device be used?
3. In what order should data be recorded?
4. How should data files be constructed in the computer?
5. How should data be checked?
6. Should data from the $98 \times 16 = 1578$ trees be analysed all at once?

As these and many other questions can be asked, this chapter is intended to provide guidance on a systematic approach to the collection and pre-processing of data before analysis of variance. We will go through the process using Genstat.

3.2 Pre-processing using Genstat

3.2.1 Data files

To facilitate the use of Genstat for this phase of the data analysis, it is important that the data be collected and entered into the computer in an appropriate manner. A lot of time can be wasted and errors made in using Genstat to read files that have been constructed without any guidelines. The rules are quite simple and should be followed as a matter of course:

1. Data sheets should be prepared with factor indexing information included. For an RCB design, factor indexing would be replicate number, plot number, tree number and seedlot number.
2. Indexing information should be in field order, i.e. the order of the plots and trees on the data sheets should represent the order in which the trees are to be measured. Thus, plots 1 and 2 would normally be adjacent to each other in the field, and so on.
3. One line on the data sheet should be used for each experimental unit. In our example a unit is a tree, so there should be one line per tree. Measurements taken on each tree, such as height and diameter, can be put in columns across the data sheet.
4. Leaving blanks in data sheets is usually not advisable. Missing measurements should be indicated by using the * symbol.
5. The content of data sheets should be entered into the computer using the same structure, thus reducing the possibility of errors in data entry.
6. If a text file is used, data should be aligned in columns with blank separators between multiple entries on a line. It is preferable, however, to use a spreadsheet such as Excel for data file construction.
7. Once a data file has been constructed, it should be checked in detail against the original data sheets. This can save a lot of time in data pre-processing and analysis.

Example 3.1

We illustrate the recommended layout for data sheets with one of the trials conducted by the Australian Centre for International Agricultural Research (ACIAR) in Queensland, Australia (Experiment 309). This was a species trial planted in 1985; survival was poor. For our example we will examine only part of the data from this experiment. Five of the species with good survival have been extracted at random, namely *Acacia*, *Angophora*, *Casuarina*, *Melaleuca* and *Petalostigma*. The first eight trees in each plot were selected and for convenience we have numbered the trees in each plot from 1 to 8. We will assume the layout given in Fig. 3.1. The data are listed in Table 3.1 and include the 4.5 year measurement of

Replicate	Seedlot	Tree							
	Melaleuca	X	X	X	X	X	X	X	X
	Casuarina	X	X	X	X	X	X	X	X
1	*Petalostigma*	X	X	X	X	X	X	X	X
	Angophora	X	X	X	X	X	X	X	X
	Acacia	X	X	X	X	X	X	X	X
	Melaleuca	X	X	X	X	X	X	X	X
	Acacia	X	X	X	X	X	X	X	X
2	*Casuarina*	X	X	X	X	X	X	X	X
	Petalostigma	X	X	X	X	X	X	X	X
	Angophora	X	X	X	X	X	X	X	X

Fig. 3.1. Layout of sample experiment (part of ACIAR Experiment 309).

Table 3.1 Part of the data from the 4.5 year measurement of ACIAR Experiment 309.

Replicate number	Plot number	Tree number	Seedlot	Ht (m)	Dgl (cm)
1	1	1	Melaleuca	5.6	20.2
1	1	2	Melaleuca	5.8	20.9
1	1	3	Melaleuca	4.9	13.2
1	1	4	Melaleuca	4.0	14.7
1	1	5	Melaleuca	5.6	23.4
1	1	6	Melaleuca	5.3	13.3
1	1	7	Melaleuca	5.5	18.2
1	1	8	Melaleuca	6.9	21.2
1	2	1	Casuarina	4.8	12.9
1	2	2	Casuarina	4.5	9.6
1	2	3	Casuarina	4.9	10.1
1	2	4	Casuarina	6.7	12.7
1	2	5	Casuarina	7.0	11.2
1	2	6	Casuarina	7.7	13.3
1	2	7	Casuarina	7.0	11.8
1	2	8	Casuarina	5.8	10.3
1	3	1	Petalostigma	3.1	5.8
1	3	2	Petalostigma	3.6	11.1
1	3	3	Petalostigma	2.8	6.4
1	3	4	Petalostigma	4.0	8.2
1	3	5	Petalostigma	3.2	8.9
1	3	6	Petalostigma	3.2	7.0
1	3	7	Petalostigma	3.3	7.3

Table 3.1 Continued

1	3	8	Petalostigma	2.1	3.0
1	4	1	Angophora	8.8	15.3
1	4	2	Angophora	7.3	11.0
1	4	3	Angophora	8.6	19.3
1	4	4	Angophora	0.9	4.2
1	4	5	Angophora	7.8	17.5
1	4	6	Angophora	6.8	11.0
1	4	7	Angophora	7.1	14.8
1	4	8	Angophora	6.8	14.0
1	5	1	Acacia	12.2	23.3
1	5	2	Acacia	11.9	27.7
1	5	3	Acacia	11.9	21.0
1	5	4	Acacia	11.6	20.4
1	5	5	Acacia	10.7	20.2
1	5	6	Acacia	12.2	18.5
1	5	7	Acacia	11.2	15.2
1	5	8	Acacia	11.7	13.2
2	1	1	Melaleuca	3.8	12.7
2	1	2	Melaleuca	5.0	19.2
2	1	3	Melaleuca	4.6	15.0
2	1	4	Melaleuca	3.7	22.2
2	1	5	Melaleuca	4.7	18.1
2	1	6	Melaleuca	4.6	20.3
2	1	7	Melaleuca	5.3	14.8
2	1	8	Melaleuca	3.8	14.1
2	2	1	Acacia	10.2	12.2
2	2	2	Acacia	10.4	22.6
2	2	3	Acacia	0.9	1.7
2	2	4	Acacia	10.6	25.8
2	2	5	Acacia	10.6	15.2
2	2	6	Acacia	10.8	34.3
2	2	7	Acacia	9.2	16.7
2	2	8	Acacia	8.6	13.9
2	3	1	Casuarina	5.5	10.8
2	3	2	Casuarina	5.1	9.7
2	3	3	Casuarina	4.1	9.6
2	3	4	Casuarina	4.0	9.6
2	3	5	Casuarina	5.4	14.4
2	3	6	Casuarina	5.2	11.5

2	3	7	Casuarina	5.5	13.2
2	3	8	Casuarina	5.0	12.2
2	4	1	Petalostigma	0.7	1.0
2	4	2	Petalostigma	3.1	7.7
2	4	3	Petalostigma	1.9	5.1
2	4	4	Petalostigma	2.9	5.0
2	4	5	Petalostigma	2.4	4.3
2	4	6	Petalostigma	3.5	6.2
2	4	7	Petalostigma	1.6	4.3
2	4	8	Petalostigma	2.3	3.9
2	5	1	Angophora	8.8	15.4
2	5	2	Angophora	5.9	12.7
2	5	3	Angophora	8.5	19.5
2	5	4	Angophora	4.1	7.5
2	5	5	Angophora	8.9	15.5
2	5	6	Angophora	9.1	16.7
2	5	7	Angophora	5.8	11.6
2	5	8	Angophora	8.4	17.8

height (*ht*) and diameter at ground level (*dgl*). The plots in Table 3.1 are in field order and also replicate by replicate. It is nearly always advisable to finish collecting data from the plots of one replicate before going to the next. Usually, collecting data replicate by replicate corresponds to the natural field order, but even where an alternative exists there is an added reason for use of this collection method, namely to confound any time effect with replicates. To explain this further, suppose that an experiment is so big that the measurement team cannot collect all data in one day. The next day a different team takes over the measurement, but has a habit of reading the measuring sticks a bit short. If one replicate had been done the first day and the other replicate the second day, the height bias due to the change of measurement persons would appear in the analysis as a between-replicate effect, and although the interpretation of this effect may be difficult, the bias of the second observer would not directly affect the comparison of seedlot means.

3.2.2 Analysis using individual tree data

Once a data file has been constructed it is possible to set up an analysis of variance using Genstat. Suppose the trial is a simple randomised complete block (RCB) design with multiple-tree plots. The blocking factors are *repl*, *plot* and *tree* and the treatment factor is just *seedlot*. The number of units is the number of trees in the trial. In order to get the correct analysis of variance table, in particular the correct error term to test seedlot effects, it is necessary to recognise that there are three strata. First there is the *repl* stratum, then the plots within each replicate (*repl.plot* stratum) and finally the trees within each plot (*repl.plot.tree* stratum). Because all the trees in each plot are from the same seedlot, the *repl.plot.tree* stratum does not contain any error information between seedlots. Hence the error term from the *repl.plot* stratum should be used for testing. A consequence of this is that the analysis of variance for testing seedlots can be performed by working with plot means instead of individual trees. In this section and the next, points for and against this alternative approach to analysis will be discussed. For the moment we will continue the analysis at the *repl.plot.tree* stratum or, in short, the *trees* level.

Example 3.1 (continued)

Table C3.1 contains a Genstat program to read the data in Table 3.1 and set up the analysis of variance for the variates *ht* and *dgl*. The analysis of variance table for *ht* is given in Table 3.2. There are the three strata, namely *repl*, *repl.plot* and *repl.plot. tree*. The term for seedlots appears in the *repl.plot* stratum and so the error term for testing seedlot differences is 5.936 on only four degrees of freedom. The number of degrees of freedom is low here, because we have chosen only five seedlots in our example and there are only two replicates. At least four replicates would normally be preferable (see Section 2.4.2).

   The plot of residuals (i.e. deviations of the observations from the fitted values) against those fitted values (Fig. 3.2) shows that there are at least two and probably

---

Table 3.2   Genstat output from analysis of variance of tree height for Example 3.1.

Analysis of variance

Variate: v[1]; ht – height (m)

Source of variation	d.f.	s.s.	m.s.	v.r.	F pr.
repl stratum	1	20.301	20.301	3.42	
repl.plot stratum					
seedlot	4	505.868	126.467	21.30	0.006
Residual	4	23.746	5.936	2.37	
repl.plot.tree stratum	70	175.614	2.509		
Total	79	725.529			

Tables of means

Variate: v[1]; ht – height (m)

Grand mean 6.12

seedlot	Acacia	Angophora	Casuarina	Melaleuca	Petalostigma
	10.29	7.10	5.51	4.94	2.73

three values that need checking. Setting three values to missing (i.e. units 28, 51 and 76) reduces the residual mean square in the *repl.plot* stratum from 5.936 to 2.022 (Table 3.3) thus highlighting the effect of unusual data on the analysis of variance table.

The analysis of data from individual trees can cause difficulties. Some points to consider are the following:

1. The number of measured trees per plot will usually vary due to mortality. In this case the analysis of variance command (**anova**) in Genstat will automatically estimate missing values for those trees that have not been measured. This inflates the seedlot mean square and is undesirable when more than a few trees are missing.
2. The detection of outliers using the plot of residuals against fitted values can be laborious and time-consuming with large data sets. It is necessary to locate and

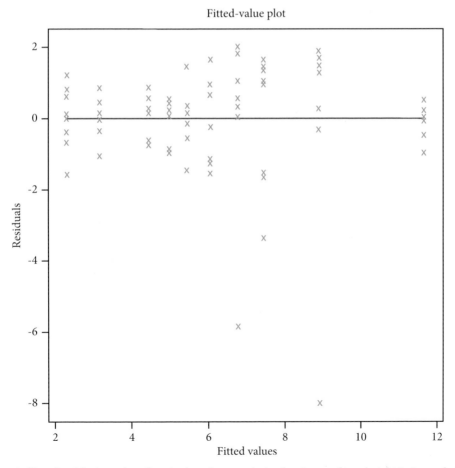

Fig. 3.2. Plot of residuals against fitted values from analysis of variance of tree height in Example 3.1.

Table 3.3 Genstat output from analysis of variance of tree height for Example 3.1 (3 values excluded).

Analysis of variance

Variate: v[1]; ht – height (m)

Source of variation	d.f.	(m.v.)	s.s.	m.s.	v.r.	F pr.
repl stratum	1		14.4743	14.4743	7.16	
repl.plot stratum						
seedlot	4		610.3762	152.5940	75.45	<.001
Residual	4		8.0894	2.0224	2.70	
repl.plot.tree stratum	67	(3)	50.2332	0.7497		
Total	76	(3)	665.0325			

Tables of means

Variate: v[1]; ht – height (m)

Grand mean 6.362

seedlot	Acacia	Angophora	Casuarina	Melaleuca	Petalostigma
	10.866	7.757	5.513	4.944	2.731

Effects of competition from adjacent trees must be avoided. This *Eucalyptus grandis* trial at Mtao, Zimbabwe (left) has been adversely affected by the large eucalypts on the right even though a trench was dug between them.

check questionable values as indicated by the residual plot and this may entail having to search lengthy data listings for particular data values. Large numbers of values tend to decrease the sensitivity of the residual plot and so it may take several analyses of variance to adequately screen the data.

3. The residual mean square in the *repl.plot.tree* stratum of Table 3.2 is derived by pooling the between-tree-within-plot variances (called plot variances) over all plots. As part of the routine checking of statistical assumptions, we should confirm that plot variances are homogeneous over the whole experiment (discussed further in the next section). If they are not, it may be necessary to modify the approach to the analysis.

An alternative general approach to analysis is given in the next section.

3.2.3   Analysis using plot summary data

The points raised at the end of the previous section are applicable to most data sets. In fact considerable time is required for the process of data screening and pre-processing, i.e. checking for possible outliers and making sure that statistical assumptions are satisfied. There is not much point obtaining an analysis of variance table if one or more of the data points are contributing excessively to the residual mean square and need to be excluded. Equally, if the plot variances differ enormously from plot to plot, it is not appropriate to construct a pooled within-plot variance such as in Table 3.2. On balance, it is much better to build up to an analysis of variance progressively involving all the data, by working in stages. This is the message underlying our whole approach to analysing data both within and across several sites.

With the analysis in Table 3.2, the first thing to recognise is that the residual for testing seedlots comes from the *repl.plot* stratum which is really the stratum that deals with plot means. In other words, the total number of degrees of freedom in the top two strata is nine, one fewer than the number of plots. Therefore we could carry out the analysis of variance for the top two strata if we had only plot means rather than individual tree data. In fact, for large experiments with multiple-tree plots, analysis of plot means is the only sensible approach. It should be pointed out, however, that when any trees are lost, the analysis of plot means is only approximate. Provided the number of trees per plot (called plot counts) is at least five, this approximate analysis is quite adequate. Otherwise, a simple analysis of plot means can still be carried out if plots with low plot counts are looked at carefully to make sure that the mean value of a small number of trees does not generate an outlier in the analysis. In any case, such a problem should show up in a graph of residuals against fitted values at the *plots* level provided there is sufficient replication in the trial. Alternatively, a weighted analysis of variance can be carried out using plot counts as weights, but it would be unusual to have to resort to such a complicated method.

Hence the idea is to operate on each plot separately and obtain some summary statistics at the *plots* level which will enable the overall analysis of variance table as given in Table 3.2 (or at least an approximate version of it) to be pieced together. The data are read in plot by plot and the plot means, counts and variances are calculated for each variate. Derived variates such as the square of tree diameter can also be summarised at the *plots* level so that plot basal area can be calculated. The plot variances are used in several ways. They provide a very effective tool for detecting unusual values in the data; an outlier will cause a large plot variance which is easily detected by scanning the vector (i.e. the list) of plot variances. We can then decide whether it is reasonable to pool the variances to form the residual mean square which appears in the bottom stratum of Table 3.2.

The vector of plot variances (or, more commonly, the logarithm of the plot variances plus one) can also be analysed to determine if the magnitude of the variances is linked in any way to particular seedlots. If there is a strong relationship between the plot means and variances, some transformation of the original data may be appropriate before analysis. Note that we usually add one to all the plot variances to overcome the difficulty of taking the logarithm of small variances less than one (giving a negative result); this adjustment has little effect on our primary purpose of comparing the plot variances for seedlots.

The vector of plot counts can be converted into a plot survival variate, i.e. plot count divided by the total number of trees originally in the plot, multiplied by 100 to get a per cent survival. This variate can be analysed at the *plots* level to investigate the survival of seedlots.

Example 3.1 (continued)

The data in Table 3.1 can be summarised to the *plots* level using the Genstat program in Table C3.2. This is a very simple program that reads data from each plot separately and for the variate *ht* calculates the plot means, variances and counts. In our example every plot has the full quota of eight trees, but in practice there are likely to be tree losses. With data files in field order it is convenient just to read the data plot by plot, carry out the calculations and then write the plot summary information onto an output file. The output from the program in Table C3.2 has been used to produce Table 3.4. Looking at the plot variances, it is clear that plot 2 in replicate 2, plot 4 in replicate 1 and probably plot 5 in replicate 2 have high variances and require closer investigation. It is then easy to go to the data listing and see what is causing the high variance. The two very low values of 0.9 for height in plot 2 in replicate 2 and plot 4 in replicate 1 stand out as being responsible for the high variances, so these should be replaced by missing values. High variances are often caused by incorrect data entry, which may be evident on

Table 3.4 Plot summary file for tree height.

Replicate number	Plot number	Seedlot	Plot mean	Plot variance	Tree count
1	1	Melaleuca	5.45	0.67	8
1	2	Casuarina	6.05	1.47	8
1	3	Petalostigma	3.16	0.31	8
1	4	Angophora	6.76	6.20	8
1	5	Acacia	11.68	0.26	8
2	1	Melaleuca	4.44	0.36	8
2	2	Acacia	8.91	11.07	8
2	3	Casuarina	4.98	0.36	8
2	4	Petalostigma	2.30	0.81	8
2	5	Angophora	7.44	3.57	8

Plots should be adequately labelled. This *Casuarina equisetifolia* trial at Ratchaburi, Thailand, has excellent permanent labels which are good for visitors and data collection alike. (Photo Khongsak Pinyopusarerk)

checking the original data sheets, but in this case it looks as though the trees in question were just very small and not representative of the particular seedlot and so it is reasonable to exclude them. Some care and judgment must be exercised when excluding data because there is always the danger of biasing the results by incorrectly dropping out data. For example, low values can sometimes provide information on the degree of selfing (Eldridge *et al.* 1993, chapter 19) or undesirable genetic effects such as runts (Barnes *et al.* 1987).

The large variance for plot 5 in replicate 2 is mainly due to one low value (4.1). It is not clear that this value should be deleted, but we would do so because the plot variance (3.57) is more than double the next biggest variance (1.47). This simple test seems a good rule of thumb. There are statistical tests that can be used to test variances, but in general we prefer to be more forgiving than would be dictated by a ratio test of variances (Snedecor and Cochran 1989, section 6.12). Once data have been screened with the aid of the plot variances, these variances can be pooled to get an overall estimate for the within-plot variance and an analysis of variance table can constructed for the plot means. In this example, however, we want to show the numerical connection between analysis of plot means and the analysis of individual tree data. Therefore, even though we know from Table 3.4 that three data points should in practice be excluded, we will continue with the full data set. The average of the plot variances in Table 3.4 would thus be 2.509, agreeing with the *repl.plot.tree* stratum residual in Table 3.2. If the plot counts are not equal, a weighted mean of plot variances strictly should be taken, but if the counts are all at least five out of eight, e.g. more than 50%, this complication can usually be overlooked.

Table 3.5 Genstat output from analysis of variance of height means for Example 3.1.

Analysis of variance

Variate: v[1]; ht – height (m)

Source of variation	d.f.	s.s.	m.s.	v.r.	F pr.
repl stratum	1	2.5301	2.5301	3.38	
repl.plot stratum					
seedlot	4	63.2595	15.8149	21.16	0.006
Residual	4	2.9899	0.7475		
Total	9	68.7794			

Tables of means

Variate: v[1]; ht – height (m)

Grand mean 6.12

seedlot	Acacia	Angophora	Casuarina	Melaleuca	Petalostigma
	10.29	7.10	5.52	4.95	2.73

The Genstat program to analyse the plot summary file in Table 3.4 is given in Table C3.3, with output for height means in Table 3.5. It is easy to see that by dividing the sums of squares and mean squares in Table 3.2 by eight (the number of trees per plot), we get the results in Table 3.5. The scaled average of the plot variances (2.509/8 = 0.314) can be manually added to Table 3.5 to get the complete analysis of variance table. Being aware of the need to scale results when carrying out analyses at different strata is very important when an overall analysis incorporating a number of sites is carried out. This is discussed further in Chapter 5.

3.2.4 Summary

The recommended method for analysing large trials with multiple-tree plots, as discussed in detail in Sections 3.2.1–3.2.3, is as follows:

Eucalyptus microtheca provenance/progeny trial at Matopos, Zimbabwe, at about one year of age. Caleb Mhongwe is demonstrating the unusual growth rate from this species at this site.

1. Index data sheets with replicate, plot, tree and seedlot information and make sure the plots are in field order.
2. Assign each field unit (normally a tree) a single line of the data sheets.
3. Record measurements on variates across the data sheets.
4. Enter data into the computer in the same format as they are recorded on the data sheets, preferably using a spreadsheet such as Excel. If a text file is used, separate entries with blank spaces.
5. Use something like the program in Table C3.2 to operate on the data file plot by plot in order to produce summary statistics for each plot, such as means, counts and variances.
6. Use plot variances to screen the data for values that need to be checked and possibly changed or excluded. Take the average of the plot variances to get an estimate of residual mean square in the *repl.plot.tree* stratum of an overall analysis of variance.
7. Analyse plot means and plot per cent survival (plot counts divided by number of trees originally in the plot, multiplied by 100). An arcsine transformation of the plot per cent survival may be advisable if there are many very high or very low percentages. We can also analyse the log-transformed (plot variances plus

Trees are grown for other purposes in addition to wood production. This is a still for processing leaves from *Eucalyptus globulus* in Spain. (Photo Robin Cromer)

one) in order to determine whether the amount of tree-to-tree variation is related to seedlot identity.

The main advantages of undertaking the analysis of variance on the plot means rather than on data for individual trees are that (i) pre-processing of the data can be more easily carried out, e.g. using the plot variances to check for unusual data values and (ii) large trials with multiple-tree plots can be handled more effectively, particularly where there has been some mortality. If, however, after data screening at the *plots* level, breeding values are required for individual trees (Chapter 6), an individual tree model can then be pursued.

Experiments with single-tree plots and many replicates cannot be reduced in size using the method mentioned above, and tree losses in such experiments can cause major problems in analysis. Another problem with single-tree plot experiments is that it is not possible to get an estimate of the *trees* level variance component (Section 6.2.1). Other points relating to single-tree plot experiments have been made in Section 2.3.

The main disadvantage of the analysis of variance at the *plots* level is that when there are tree losses the analysis is only approximate, but it should be borne in mind that the Genstat analysis of variance of individual tree data is also only approximate when there are missing values. When the number of trees per plot does not drop too low, the approximate analysis is perfectly acceptable. Alternatively, a weighted analysis of variance can be carried out using the plot counts as weights.

3.3 Pre-processing using Excel

It is common practice to hold the data from experiments in Microsoft Excel, or similar spreadsheet software programs. It would be very tedious to manually construct an indexed spreadsheet for a large experiment with multiple blocking factors and many treatments. Any mistakes in spreadsheet indexing will cause errors in subsequent analysis.

The second edition of this book described a purpose-designed data pre-processing software package, DataPlus, which constructs appropriately structured and indexed Excel spreadsheets matching a wide range of experimental designs. DataPlus generates Excel spreadsheets in the format that we have advocated in Section 3.2.1, and as shown in Table 3.1. Some researchers continue to use DataPlus, but this package is no longer supported, and is not available to new users.

Excel spreadsheets can be constructed using the software package CycDesigN (Section 7.8). The type of design, its dimensions and the list of treatments are specified and a randomised design is produced by CycDesigN. The program outputs an Excel file suitable for data collection, with trees in field order, indexed by the blocking and treatment factors, as in Table 3.1.

It is recommended that separate worksheets of the Excel data file are used to record detailed maps of the experiment, showing the spatial layout of all field and treatment factors, and indicating the correct sequence of measurement, both between and within plots; this information must be accessible to the field measuring team. For those working with experiments that have not been designed using CycDesigN, it is possible to set up a correctly dimensioned Excel file for data collection and analysis using commands within Excel. For example, the Excel command *vlookup* can be used to correctly allocate treatment identities to individual tree rows in the data collection file by interrogating a summary two-way table of treatment allocations to the different replicates.

Many researchers use an electronic field data capture device, which can directly import the indexed Excel worksheet and display for data entry a moving window of indexed cells in field order. Others will print out the indexed Excel spreadsheet, page by page in field order, with column and row indexing shown on each printed page, for manual data entry.

If the data has been recorded manually onto paper sheets in the field, it must later be transferred to the indexed spreadsheet. This is best done by one worker calling the data values, row by row (for example height (*ht*), diameter at breast

Table 3.6 Procedures for data checking and cleaning using Microsoft Excel.

Excel command/action	Check
Data Filter	*Data Filter* is engaged and each column is examined in turn. Alpha-numeric values in the selected column will be shown in ascending order. In Example 3.1, the column headed Replicate number should only show the numbers 1 and 2, while the column headed *Ht* should show only the numerical values in the anticipated height range and * for missing trees. Any values outside the anticipated range (e.g. a mis-typed 56, 0.56, 5..6 or 5. 6 instead of 5.6, for the first tree in Table 3.1) can easily be identified and corrected as necessary.
Pivot Table	Pivot tables can be created (on separate Excel worksheets) to check the data in various ways. Setting up a pivot table with replicate as a *Column*, plot as a *Row* and seedlot as a *Value*, provides a check that a randomised complete block design is correctly dimensioned and indexed. Another pivot table with replicate as a *Column*, seedlot as a *Row* and a data variate such as height as a *Value* in the pivot table checks that all treatments are present in all replicates. Using different settings for *Value Field Settings* for height, we can check the number of entries in each cell of the two-way replicate x treatment table (*Count*), the mean value for each cell (*Average*), the variance (*Variance*), and the number of non-missing trees (*Countnum*). If the *Countnum* replicate x treatment pivot table is copied to a different area of the worksheet, a two-way table displaying percentage survival of each treatment in each replicate can be calculated simply from the copy.
	If the individual trees in the trial are indexed according to an overall two-dimensional x-y structure, an x-y pivot table can be created for a variate of interest, such as tree height, and 'heat maps' can be created by highlighting a copy of this pivot table and selecting *Conditional Formatting* to display level intervals as different colours. This can provide a quick spatial visualisation – for example the spatial distribution of tree size classes or disease symptoms. Such visualisations may suggest the importance of blocking factors such as long columns for field variation (Section 7.5.1), and they provide a simple point of entry for formal spatial analysis (Section 7.6).
Creating and examining biplots	It is straightforward to create x-y scatter plots for pairs of variates in Excel; for example we expect *ht* and *dbh* to be strongly correlated. Outliers can then be investigated.
Constructing and checking additional data columns	We can create additional data columns and populate them using arithmetic formulas in Excel, for example, the column '*dbh* increment', calculated as *dbh*2022 – *dbh*2021. Here, if any tree returns a negative value (quickly identified using the *Data Filter* command) or if a tree's *dbh* increases at a rate that is biologically implausible, it must be investigated. If missing values are entered as * as recommended, trees that are missing at either or both measures will return a value of ##### and can similarly be checked. Another common practice is to use Excel to calculate, for subsequent statistical analysis, additional variates derived from the measured variates. For example, stem basal area can be calculated from *dbh*, while stem volume is often calculated from *dbh*, *ht* and a form factor.

height (*dbh*) etc. of each tree), with regular checks on of row identities, and another worker typing the called values into the Excel spreadsheet. Once the data has been typed in, it is read back from Excel, with another worker checking that all data values match back to those in the field sheets. If only one worker is available for data entry, read-back and checking is still a necessary step. Our experience is that there will inevitably be data entry errors; many more if read-back is not carried out. Failure to check data entry often results in faulty statistical analysis and false conclusions.

Typically, a field trial will be measured two or more times. The values for a variate from the previous measure, for example tree height, should be displayed immediately to the left of the data variates that are to be collected in subsequent measures. This provides an in-field check that data collection is proceeding correctly. Missing trees should show up on the same rows as the previous year, and trees that were unusual at the previous measure (e.g. very large or very small) should be similarly recognisable.

Excel can be used to carry out initial checking and pre-processing of the data, once correctly entered. It is recommended practice to first identify the Excel worksheet carrying the data as entered from the field with a name such as 'field data'. Then copy the entire data set to a separate worksheet, giving this new worksheet a separate name such as 'checking data'. Use this worksheet to carry out various data checks and mark up any changes. A clean data set can then be copied to another worksheet named as 'data for analysis'. This worksheet can be locked to prevent accidental changes, and back-up copies of the Excel file made for safe keeping.

The checking methods in Excel given in Table 3.6 are particularly useful (see chapter 15 of White *et al.* (2007) for further discussion of data checking and cleaning). Using Excel for data checking and pre-processing will give researchers a good 'feel' for their data. Many problems can be identified and corrected before proceeding to formal statistical analysis.

Once a data set has been checked in Excel, it can be used as input for a statistical package.

4

Experimental design

4.1   Introduction

In Sections 2.6.1 and 2.6.2 analysis of variance tables for completely randomised and randomised complete block (RCB) designs were discussed. These designs were introduced to demonstrate the concept of strata, which is fundamental to understanding experimental design and in particular the structure of the analysis of variance table.

This chapter examines in more detail completely randomised and RCB designs as well as factorial and split-plot designs. We introduce the idea of a linear model, the concept underlying the analysis of variance table in experimental design. Following the arguments presented in Chapter 3, all operations here will be carried out at the *plots* level, i.e. if a trial has multiple-tree plots we assume that a plot summary file has been created and this is the data file to be analysed.

The types of experimental design covered in this chapter are not exhaustive, but they do include most of the complete block design types likely to be used in forestry field trials. Chapters 7 and 8 will concentrate on incomplete block designs, which are more complicated but necessary for trials involving many treatments.

4.2   Simple designs

An experiment is analysed by assuming a model for the data and then testing to see how well the model fits the data. For the experimental designs in this book, the assumed model consists of the sum of a number of components. This is known as a linear model. The components in the linear model contain parameters which have to be estimated using a method called 'least squares'. The theory of least-squares estimation is described in the book by Scheffé (1959).

4.2.1 Completely randomised design

Suppose there are v treatments each with r replications and the treatments are assigned at random to $n = vr$ plots. As discussed in Chapter 2, at the *plots* level there would be just one stratum for this completely randomised design.

We assume a model of the following form:

$$(\text{observation}) = (\text{overall mean}) + (\text{treatment effect}) + (\text{residual})$$

The interpretation of this model is that an observation (normally a plot mean) consists of an overall mean plus a contribution due to the particular treatment. The treatment effect is a deviation about the overall mean, i.e. a good treatment will have a positive deviation and a poor treatment will have a negative deviation. The overall mean and treatment effects are known as parameters in the linear model and they have to be estimated from the data, namely the observations from the vr plots. The linear model can be written symbolically as:

$$Y_{ij} = \mu + \tau_j + \varepsilon_{ij}$$

where the Y_{ij} ($i = 1,2,\ldots,r; j = 1,2,\ldots,v$) are the observations; μ is a parameter for the overall mean; the τ_j are parameters for the v treatments and the ε_{ij} are the residuals in the model, i.e. what is left over from the proposed model. By making the assumption that the ε_{ij} are distributed according to the normal distribution with zero mean and variance σ^2, it is possible to test the significance of treatment differences. The analysis of variance table provides the quantities needed to carry out this test, known as the F-test or variance ratio test of significance. We will not treat these concepts in general but merely demonstrate them via an example.

Example 4.1

Look again at Example 2.1, a completely randomised design for two seedlots (SO and P) and replication of eight. The data are in Table 2.1.

The least squares estimator ($\hat{\mu}$) for the overall mean is simply the average of all 16 plot means, namely:

$$\hat{\mu} = (30.38 + 28.16 + \ldots + 28.34)/16$$
$$= 29.637$$

The estimated seedlot effects are the deviations of the means for each seedlot from the overall mean. So if $\hat{\tau}_1$ is the estimated effect for SO and $\hat{\tau}_2$ is the estimated effect for P, then:

$$\hat{\tau}_1 = (30.38 + 27.91 + \ldots + 33.53)/8 - 29.637$$
$$= 30.581 - 29.637$$
$$= 0.944$$

and:

$$\hat{\tau}_2 = (28.16 + 25.62 + \ldots + 28.34)/8 - 29.637$$
$$= 28.693 - 29.637$$
$$= -0.944$$

Once the parameters have been estimated, we can arrive at the fitted values from the model, namely $\hat{\mu} + \hat{\tau}_1 = 30.581$ for seedlot SO and $\hat{\mu} + \hat{\tau}_2 = 28.693$ for seedlot P. The difference between the observations and the fitted values gives the residuals:

$$\varepsilon_{11} = Y_{11} - (\hat{\mu} + \hat{\tau}_1) = 30.38 - 30.581 = -0.201$$
$$\varepsilon_{12} = Y_{12} - (\hat{\mu} + \hat{\tau}_2) = 28.16 - 28.693 = -0.533$$
$$\vdots$$
$$\vdots$$
$$\varepsilon_{81} = Y_{81} - (\hat{\mu} + \hat{\tau}_1) = 33.53 - 30.581 = -2.949$$
$$\varepsilon_{82} = Y_{82} - (\hat{\mu} + \hat{\tau}_2) = 28.34 - 28.693 = -0.353$$

The residuals are usually plotted against the fitted values as a data screening mechanism. We discussed this type of plot in Example 3.1 and Fig. 3.2 when looking at analysis of variance at the *trees* level. As emphasised in Chapter 3, however, we recommend construction of a plot summary file and analysis at the *plots* level. Then the plot of residuals against the fitted values will be first encountered at the *plots* level.

If the assumed linear model is a good approximation to the observations Y_{ij} then the residuals ε_{ij} will be small. In order to quantify the magnitude of the ε_{ij} relative to differences between the seedlot effects, we construct an analysis of variance table as done in Table 2.2. In this table the seedlot sum of squares is calculated as $8(\hat{\tau}_1^2 + \hat{\tau}_2^2) = 14.258$ where the coefficient 8 refers to the number of replications of each seedlot. Note the discrepancy between this and the value given in Table 2.2 (14.270). The difference is due to rounding error when we obtained our estimates for $\hat{\tau}_1$ and $\hat{\tau}_2$. It is well known that the above method for calculating sums of squares is not accurate, but our interest here is to demonstrate the derivation of the quantities in Table 2.2; we will rely on the computer to do more exact calculations.

The residual sum of squares is obtained as $\varepsilon_{11}^2 + \varepsilon_{12}^2 + \ldots + \varepsilon_{82}^2 = 61.410$. Each sum of squares is divided by the number of degrees of freedom (d.f.) to give the seedlot and residual mean squares as in Table 2.2. The ratio of the seedlot mean square to the residual mean square (3.25) is distributed according to the F-distribution, if we make the assumption that the ε_{ij} are normally and independently distributed with zero mean and variance σ^2. The residual mean square (s^2) is an unbiased estimator for σ^2. The assumption that residuals are normally distributed should be checked in each case. The plot of residuals against fitted values will check if the assumption is violated by the presence of outliers, or if there is a relationship between the fitted values and the spread of the residuals, thereby necessitating a variance-stabilising transformation or the specification of an alternative model.

For multiple-tree plots, analysis of variance of the log-transformed (plot variances plus one) provides a convenient method for checking the assumption that all ε_{ij} have equal variance. Often this analysis will show significant seedlot differences, thus indicating a relationship between plot variances and seedlots. Once again, some transformation or change of model is warranted in order to satisfy the assumptions made.

4.2.2 Randomised complete block design

Here we have v treatments in r replicates. As discussed in Section 2.6.2, there are two strata and we assume a model of the form:

$$\left(\text{observation}\right) = \left(\text{overall mean}\right) + \left(\text{replicate effect}\right) + \left(\text{treatment effect}\right) + \left(\text{residual}\right)$$

or symbolically:

$$Y_{ij} = \mu + \rho_i + \tau_j + \varepsilon_{ij}$$

where the ρ_i ($i = 1,2,...,r$) are parameters for the r replicates. By incorporating parameters for replicate effects in the model we aim to reduce the size of the residuals ε_{ij}, thereby reducing the residual mean square in the analysis of variance table.

Example 4.2

Look again at the RCB design discussed in Example 2.2 and presented in Fig. 2.4. The estimators $\hat{\mu}$, $\hat{\tau}_1$ and $\hat{\tau}_2$ remain the same as for the completely randomised design in Example 4.1. The least squares estimators for replicates are obtained as follows:

$$\hat{\rho}_1 = (30.38 + 28.16)/2 - 29.637$$
$$= 29.27 - 29.637$$
$$= -0.367$$
$$\vdots$$
$$\vdots$$
$$\hat{\rho}_8 = (28.34 + 33.53)/2 - 29.637$$
$$= 30.935 - 29.637$$
$$= 1.298$$

The fitted values from the model have the form:

$$\hat{Y}_{ij} = \hat{\mu} + \hat{\rho}_i + \hat{\tau}_j$$

and so the residuals become:

$$\varepsilon_{11} = Y_{11} - \hat{Y}_{11} = 30.38 - 30.214 = 0.166$$

$$\varepsilon_{12} = Y_{12} - \hat{Y}_{12} = 28.16 - 28.326 = -0.166$$

$$\vdots$$

$$\vdots$$

$$\varepsilon_{81} = Y_{81} - \hat{Y}_{81} = 33.53 - 31.879 = 1.651$$

$$\varepsilon_{82} = Y_{82} - \hat{Y}_{82} = 28.34 - 29.991 = -1.651$$

The replicate sum of squares is equal to $2(\hat{\rho}_1^2 + \hat{\rho}_2^2 + \ldots + \hat{\rho}_8^2) = 48.867$ where the coefficient 2 is the number of plots in each replicate (Table 2.3). The residual sum of squares is $\varepsilon_{11}^2 + \varepsilon_{12}^2 + \ldots + \varepsilon_{82}^2$ but now the ε_{ij} are smaller than for the completely randomised design, a result of including replicates in the model. The sum of squares is 12.543 and is only $(12.543/61.410) \times 100 = 20.4\%$ of that for the completely randomised design. Hence including replicates in the model has considerably improved the approximation of the values derived from the model to the data. In other words, this model is a better fit to the data than the previous model.

As shown in Table 2.3, mean squares can be calculated for replicates, seedlots and the residual, giving a variance ratio of 7.96 for seedlots. We test the null hypothesis that seedlots do not differ from each other by comparing this variance ratio with tables of the F-distribution for 1 and 7 d.f. The variance ratio (7.96) is larger than the tabulated F-value (5.59) at the 5% level of significance so we reject the null hypothesis and accept that there are significant differences between the two seedlots.

The standard error for the difference between two seedlot means is $\sqrt{2s^2/8} = 0.67$, where $s^2 = 1.792$. In this case, where a pairwise comparison has been chosen in advance, the seedlots may be compared using a t-test. The value of t for the difference between SO and P is $1.89/0.67 = 2.82$ compared with the t-value of 2.365 for 7 d.f. at the 5% probability level, i.e. a significant difference. For this example, where there are only two seedlots under test, the t-test is exactly comparable with the F-test as the critical points for the t-distribution are the square roots of the corresponding F-distribution values (i.e. $2.365 = \sqrt{5.59}$). For more than two seedlots, we can calculate a least significant difference (LSD) for the seedlot means. This is the product of the critical value of t and the standard error for the difference between two seedlot means. In this case the LSD at the 5%

Provenance trial of *Acacia mangium* growing near Guangzhou, China. Bulked provenance seedlots were used in this collaborative trial between the Research Institute of Tropical Forestry (Chinese Academy of Forestry) and ACIAR (Project 8848).

probability level is $2.365 \times 0.67 = 1.585$ which is smaller than the difference between the two seedlot means (1.89) and is another way to show the significant result.

4.3   Factorial designs

So far we have considered experimental designs with only one treatment factor. It is very common, however, to conduct field trials or glasshouse experiments which involve looking at differences between treatment levels in combination with another treatment factor, such as different irrigation levels. This would be a two-factor experiment. Another example would be a fertiliser experiment involving levels of nitrogen and of phosphorus on a single seedlot; if different seedlots are used this would become a three-factor experiment. It is not very common to have many treatment factors in forestry trials, but in other contexts, such as industrial or laboratory experiments, multi-factor designs are used. These can lead to considerations of confounding and aliasing (Cochran and Cox 1957, chapters 6, 6A). We will not pursue this branch of experimental design. Instead we will look in more detail at a two-factor experiment involving treatment factors A and B with a and b levels respectively and r replicates.

The linear model is of the form:

$$(\text{observation}) = (\text{overall mean}) + (\text{replicate effect}) + (A \text{ effect}) +$$
$$(B \text{ effect}) + (A.B \text{ effect}) + (\text{residual})$$

or symbolically:

$$Y_{ijk} = \mu + \rho_i + A_j + B_k + A.B_{jk} + \varepsilon_{ijk} \tag{4.1}$$

The $A.B_{jk}$ ($j = 1,2,...,a$; $k = 1,2,...,b$) represent the interaction between the treatment factors A and B. These parameters allow us to accommodate the situation where the behaviour of factor B differs at different levels of A. One of the strengths of factorial experiments is that they provide information on treatment interaction which cannot be derived from estimated means for the treatment factors taken individually.

Example 4.3

Major seed distributors, such as the Australian Tree Seed Centre (ATSC), routinely conduct seed viability tests so that, when seed is dispatched, the purchaser has an indication of the germination percentage of the seed. Other areas of investigation include storage conditions, loss of viability and methods of pre-treatment for seed, so the ATSC can give seed purchasers helpful and accurate information on suitable germination procedures and viability.

As part of the ATSC research program, a series of experiments was conducted in 1992 by Debbie Solomon on provenances of *Acacia mangium* to investigate methods of pre-treatment and loss of viability of stored seed. Each experiment involved six seedlots of *Acacia mangium* and four seed pre-treatments in a factorial design with three replicates. The seedlots were CSIRO numbers 18265, 18249, 18248, 18211, 18212 and 18217. The pre-treatments on the seed were control, nicking, boiling water followed by soaking, and boiling water for one minute. For each replicate and treatment combination, 25 seeds were placed in a germination dish and the variate recorded was the number of germinants (out of 25) after a period of 21 days. Here we concentrate on the third of these experiments, conducted in August 1992.

Petri dishes of seeds in a germination cabinet. (Photo Kron Aken)

The 72 dishes were laid out in an experimental design within a germination cabinet so that each replicate represented a shelf in the cabinet. The 24 dishes on each shelf were arranged so that possible variation within the cabinet would be assessed and separated from variation due to the two treatment factors. In other words, in addition to the blocking factor (*repl*) to adjust for possible variation between shelves, there were two other blocking factors (*row* and *column*) within each shelf. The row and column blocking factors complicate the analysis and will be discussed in Chapters 7 and 8. The experimental layout and germination counts are given in Fig. 4.1.

The Genstat specification for this experiment is:

block   repl / (row.column)

treatment   treat * seedlot

Row	Column 1	Column 2	Column 3	Column 4	
1	2 4 (15)	3 2 (17)	1 6 (2)	4 5 (23)	
2	1 5 (0)	4 4 (24)	2 2 (18)	3 6 (10)	
3	4 1 (24)	1 3 (0)	2 5 (13)	3 1 (19)	Replicate 1
4	3 3 (14)	2 6 (1)	1 1 (21)	4 2 (0)	
5	1 2 (0)	3 5 (15)	4 3 (16)	2 3 (11)	
6	4 6 (16)	2 1 (20)	3 4 (21)	1 4 (0)	
1	3 5 (20)	1 1 (0)	4 1 (18)	2 4 (20)	
2	3 2 (18)	4 4 (19)	2 3 (9)	1 6 (2)	
3	4 3 (17)	2 5 (15)	1 5 (1)	3 1 (17)	Replicate 2
4	1 4 (0)	3 6 (14)	2 6 (14)	4 2 (19)	
5	4 6 (20)	2 2 (6)	3 4 (19)	1 3 (2)	
6	2 1 (18)	3 3 (15)	1 2 (0)	4 5 (21)	
1	4 3 (18)	2 5 (13)	3 3 (10)	1 6 (1)	
2	1 5 (0)	2 1 (18)	4 1 (20)	3 5 (16)	
3	2 4 (14)	3 4 (19)	4 5 (20)	1 2 (0)	Replicate 3
4	3 6 (17)	4 6 (22)	1 4 (0)	2 3 (10)	
5	1 1 (0)	4 2 (23)	2 6 (15)	3 1 (24)	
6	3 2 (11)	1 3 (0)	2 2 (13)	4 4 (23)	

Treatment combinations

	Pre-treatment		Seedlot
1	Control	1	18265
2	Nick	2	18249
3	Boiling water and soak	3	18248
4	Boiling water for 1 min	4	18211
		5	18212
		6	18217

Fig. 4.1. Layout of dishes for Example 4.3 with treatment combinations and germination counts (in brackets).

Using the * between the two treatment factors requests Genstat to include the interaction, so the resulting analysis of variance will be consistent with the linear model (4.1). There are two strata in this experiment, *repl* and *repl.row.column*. If we had considered rows and columns, the Genstat blocking structure would have been:

block   repl / (row * column)

representing four strata, *repl*, *repl.row*, *repl.column* and *repl.row.column*. The analysis using these strata is more easily carried out using the **fit** command of Genstat. A description of **fit** is given in Appendix A and further discussion of this type of analysis, known as non-orthogonal analysis of variance, is in Chapter 8.

It is readily seen from Fig. 4.1 that the germination was very poor for the control (no pre-treatment). In the analysis of variance, therefore, it is better to partition the difference between the pre-treatments into two components: (i) a comparison between the control and the three genuine pre-treatments and (ii) comparisons between the genuine pre-treatments. This type of partition is easily accomplished in Genstat by defining a new factor *contcomp* (short for 'control comparison') with two levels and using the **calculate** command to derive the values for *contcomp* (Appendix A). The Genstat code for calculation of percentage germination, *contcomp* and the analysis of variance is given in Table C4.1.

The analysis of variance of the percentage germination is presented in Table 4.1. There is clearly a very large effect for the comparison between the control and pre-treatments (*contcomp*), which we knew already. In addition, there are highly significant differences between the genuine pre-treatments (*contcomp.treat*). When we look at the plot of residuals against fitted values (Fig. 4.2), it is clear that the variation for the control pre-treatment is much less than for the other pre-treatments. This is consistent with data that are proportions, and various actions can be taken to correct this contradiction of one of the assumptions we have made about the data, namely that the residuals all have the same variance. The most common action is to make a variance-stabilising transformation such as the angular transformation (Snedecor and Cochran 1989, section 15.12). Because we are not so interested in the control pre-treatment, however, we can simply remove it from the analysis using the **restrict** command in Genstat (Appendix A) as illustrated in Table C4.2, leading to the analysis of variance in Table 4.2. In this case it is probably better to base conclusions on that analysis rather than to adjust Table 4.1 to satisfy the analysis of variance assumptions.

Table 4.2 reveals highly significant differences between the three genuine pre-treatments, with the immersion in boiling water for one minute giving better results than quick immersion in boiling water and overnight soaking, or nicking the seed. The standard error for the difference between two pre-treatment means is 3.78, leading to a least significant difference of 3.78 $t_{34} = 7.67$ at the 5% probability

Table 4.1 Genstat output from analysis of variance of percentage germination for Example 4.3.

Analysis of variance

Variate: v[1]; percent – percent=(count/25)*100

Source of variation	d.f.	s.s.	m.s.	v.r.	F pr.
repl stratum	2	35.11	17.56	0.18	
repl.row.column stratum					
contcomp	1	58542.30	58542.30	601.52	<.001
seedlot	5	2894.44	578.89	5.95	<.001
contcomp.treat	2	5300.15	2650.07	27.23	<.001
contcomp.seedlot	5	1347.04	269.41	2.77	0.029
contcomp.treat.seedlot	10	961.19	96.12	0.99	0.467
Residual	46	4476.89	97.32		
Total	71	73557.11			

Tables of means

Variate: v[1]; percent – percent=(count/25)*100

Grand mean 51.4

contcomp	1	2
	2.0	67.9
rep.	18	54

seedlot	18211	18212	18217	18248	18249	18265
	58.0	52.3	49.0	40.7	48.7	59.7

contcomp	treat	control	nick	bw&s	bw1min
1		2.0			
2			56.9	65.8	80.9

contcomp	seedlot	18211	18212	18217	18248	18249	18265
1		0.0	1.3	6.7	2.7	0.0	1.3
	rep.	3	3	3	3	3	3
2		77.3	69.3	63.1	53.3	64.9	79.1
	rep.	9	9	9	9	9	9

contcomp	treat	seedlot	18211	18212	18217	18248	18249
1	control		0.0	1.3	6.7	2.7	0.0
2	nick		65.3	54.7	57.3	40.0	49.3
	bw&s		78.7	68.0	54.7	52.0	61.3
	bw1min		88.0	85.3	77.3	68.0	84.0

Table 4.1 Continued

contcomp	treat	seedlot	18265
1	control		1.3
2	nick		74.7
	bw&s		80.0
	bw1min		82.7

Standard errors of differences of means

Table	contcomp	seedlot	contcomp treat	contcomp seedlot	
rep.	unequal	12	18	unequal	
d.f.	46	46	46	46	
s.e.d.				8.05	min.rep
	2.68	4.03	3.29	6.58	max-min
				4.65	max.rep

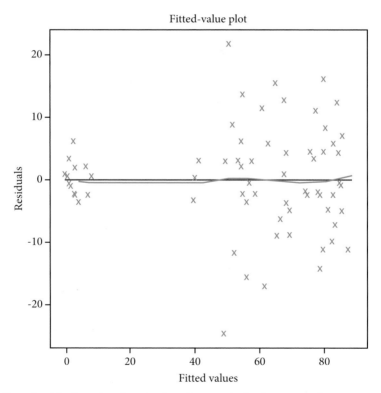

Fitted-value plot

Fig. 4.2. Plot of residuals against fitted values from analysis of variance of germination percentage for Example 4.3.

Table 4.2   Genstat output from analysis of variance of percentage germination for Example 4.3 with the control pre-treatment deleted.

Analysis of variance

Variate: v[1]; percent – percent=(count/25)*100

Source of variation	d.f.	s.s.	m.s.	v.r.	F pr.
repl stratum	2	64.6	32.3	0.25	
repl.row.column stratum					
treat	2	5300.1	2650.1	20.61	<.001
seedlot	5	4148.1	829.6	6.45	<.001
treat.seedlot	10	961.2	96.1	0.75	0.676
Residual	34	4372.7	128.6		
Total	53	14846.8			

Tables of means

Variate: v[1]; percent – percent=(count/25)*100

Grand mean 67.9

treat	control	nick	bw&s	bw1min
	56.9	65.8	80.9	

seedlot	18211	18212	18217	18248	18249	18265
	77.3	69.3	63.1	53.3	64.9	79.1

treat seedlot	18211	18212	18217	18248	18249	18265
nick	65.3	54.7	57.3	40.0	49.3	74.7
bw&s	78.7	68.0	54.7	52.0	61.3	80.0
bw1min	88.0	85.3	77.3	68.0	84.0	82.7

Standard errors of differences of means

Table	treat	seedlot	treat seedlot
rep.	18	9	3
d.f.	34	34	34
s.e.d.	3.78	5.35	9.26

level. Interestingly, there are also highly significant differences between the provenances of *Acacia mangium*, but the interaction between pre-treatments and seedlots is not significant.

4.4 Split-plot designs

Example 4.3 was a two-factor experiment where both treatment factors were 'applied' at the same stratum level, namely the *repl.row.column* stratum. Sometimes it is desirable to impose treatment factors at different stratum levels, e.g. when some treatments have to be applied on a broader scale than others. Experimental structures of this type are known as split-plot designs. There is some traditional terminology associated with split-plot designs; the blocking structures are called replicates, main-plots and sub-plots, leading to the strata *repl*, *repl.mainpl* and *repl.mainpl.subpl*. As an example the broad-scale treatment, fertiliser say, could be applied to main-plots and seedlots assigned to the sub-plots. A consequence of having treatment factors at different stratum levels is that there will be separate error terms for the factors. The model for a split-plot design is of the form:

$$(\text{observation}) = (\text{overall mean}) + (\text{replicate effect}) + (A\,\text{effect}) +$$
$$(\text{main-plot residual}) + (B\,\text{effect}) +$$
$$(A.B\,\text{effect}) + (\text{sub-plot residual})$$

or symbolically:

$$Y_{ijk} = \mu + \rho_i + A_j + \eta_{ij} + B_k + A.B_{jk} + \varepsilon_{ijk} \tag{4.2}$$

This is a split-plot experiment (aged 10 months) planted on a saline site near Hyderabad, Pakistan, in which the main plots were combinations of nutrient and fertiliser. Tree species forming sub-plots were *Conocarpus lancifolius* (left), *Acacia nilotica* (centre) and *A. ampliceps* (right). (Photo John Turnbull)

The η_{ij} ($i = 1,2,...,r$; $j = 1,2,...,a$) are the main-plot residuals and provide an extra term in the model compared with (4.1).

One advantage of split-plot designs is the extra flexibility provided by the ability to apply one treatment factor at the *repl.mainpl* stratum. A disadvantage is that comparisons between the treatment factor A parameters are made with less precision than for treatment factor B or the A.B interaction parameters.

Example 4.4

An experiment supported by the Shell Company was planted at Toolara Forest Reserve near Gympie, Queensland, in February 1987 to study the effects of irrigation and fertiliser on four seedlots of *Eucalyptus grandis*. Because of the difficulty in applying the irrigation and fertiliser treatments individually to each 7 × 6 plot of trees, the experiment was designed as a split-plot, with main-plot treatments, *irrig* and *fert*. There were two replicates of four main-plots each with four sub-plots. The experimental layout is given in Fig. 4.3, together with sub-plot means for 34-month tree height (m).

The Genstat specification for the analysis of variance is:

block   repl / mainpl / subpl

treatment   irrig * fert * seedlot

and the full program is given in Table C4.3. The output (Table 4.3) shows a very large difference between the application or non-application of fertiliser; there

			Replicate					
			1				2	
irrigation:	none	none	plus	plus	plus	plus	none	none
fertiliser:	none	plus	plus	none	none	plus	none	plus
	4	2	1	3	2	1	4	3
	4.71	16.36	14.38	4.66	5.41	14.60	4.32	14.98
	3	1	2	4	3	4	1	2
	6.23	15.29	16.89	4.95	5.73	12.21	4.16	15.98
	2	3	4	1	4	2	3	1
	7.46	13.99	11.25	5.81	5.80	14.84	5.02	14.40
	1	4	3	2	1	3	2	4
	6.39	11.08	15.58	7.50	6.39	15.00	6.79	11.98

Seedlots   1   Bulahdelah
               2   Coffs Harbour seed orchard
               3   Pomona plantation
               4   Atherton

Fig. 4.3. Layout of plots for Example 4.4 with seedlot numbers and height means.

Table 4.3 Genstat output from analysis of variance of height means for Example 4.4.

Analysis of variance

Variate: v[1]; ht – height (m)

Source of variation	d.f.	s.s.	m.s.	v.r.	F pr.
repl stratum	1	0.7564	0.7564	1.08	
repl.mainpl stratum					
irrig	1	0.1081	0.1081	0.15	0.721
fert	1	590.6485	590.6485	841.11	<.001
irrig.fert	1	0.0072	0.0072	0.01	0.926
Residual	3	2.1067	0.7022	1.05	
repl.mainpl.subpl stratum					
seedlot	3	39.6538	13.2179	19.68	<.001
irrig.seedlot	3	1.1098	0.3699	0.55	0.657
fert.seedlot	3	9.9503	3.3168	4.94	0.018
irrig.fert.seedlot	3	1.7360	0.5787	0.86	0.487
Residual	12	8.0596	0.6716		
Total	31	654.1364			

Tables of means

Variate: v[1]; ht – height (m)

Grand mean 10.00

irrig	none	plus
	9.95	10.06

fert	none	plus
	5.71	14.30

seedlot	Atherton	Bulahdelah	Coffs SO	Pomona pltn
	8.29	10.18	11.40	10.15

irrig	fert	none	plus
none		5.64	14.26
plus		5.78	14.34

irrig	seedlot	Atherton	Bulahdelah	Coffs SO	Pomona pltn
none		8.02	10.06	11.65	10.06
plus		8.55	10.30	11.16	10.24

fert	seedlot	Atherton	Bulahdelah	Coffs SO	Pomona pltn
none		4.95	5.69	6.79	5.41
plus		11.63	14.67	16.02	14.89

irrig	fert	seedlot	Atherton	Bulahdelah	Coffs SO	Pomona pltn
none	none		4.52	5.28	7.13	5.63
	plus		11.53	14.84	16.17	14.49
plus	none		5.38	6.10	6.46	5.20
	plus		11.73	14.49	15.86	15.29

Standard errors of differences of means

Table	irrig	fert	seedlot	irrig fert
rep.	16	16	8	8
d.f	3	3	12	3
s.e.d.	0.296	0.296	0.410	0.419

Table	irrig seedlot	fert seedlot	irrig fert seedlot
rep	4	4	2
s.e.d.	0.583	0.583	0.824
d.f.	14.69	14.69	14.69

are also highly significant differences between seedlots. Notice how Genstat assigns the treatment factors to the correct strata in Table 4.3. The main-plot treatment factors and their interaction appear in the *repl.mainpl* stratum while *seedlot* and the interactions with *seedlot* are all in the *repl.mainpl. subpl* stratum.

Usually the main-plot residual mean square will be higher than the sub-plot residual mean square, but in this case they are comparable. Also, with the fertilised main-plots growing so much faster than the unfertilised we would expect to have to make a variance-stabilising transformation before analysis. This is because the variance is often related to the mean, hence violating one linear model assumption. The plot of residuals against fitted values in Fig. 4.4, however, shows that in this experiment a transformation is not needed.

The seedlot-by-fertiliser interaction is significant ($P < 0.05$), but the seedlot-by-fertiliser table of means (Table 4.3) clearly shows that the interaction has been generated by the relatively poor performance of the Atherton seedlot with fertiliser and so is not worth further analysis. The standard error for the difference between two seedlot means is 0.410, leading to an LSD of $0.410\,t_{12} = 0.89$. It is clear that the Coffs Harbour seed orchard material has performed much better than the other seedlots.

Study of the design layout in Fig. 4.3 will reveal that the assignment of seedlot numbers to sub-plots has not been done at random for each main-plot. The

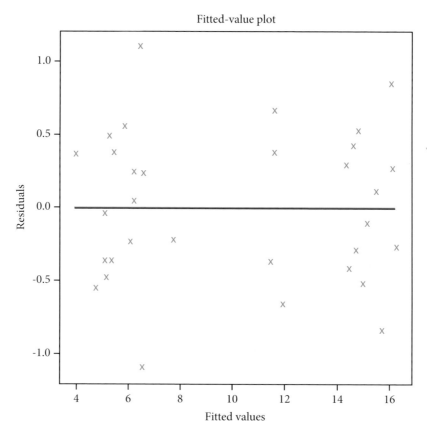

Fig. 4.4. Plot of residuals against fitted values from analysis of variance of height means for Example 4.4.

randomisation has been restricted so that within each replicate the rows of sub-plots (running across main-plots) each contain the four seedlot numbers exactly once. This is a design feature and produces an extra stratum for removing field trend in a direction at right angles to the main-plots. As we did for Example 4.3, we will ignore this extra blocking factor (*row*) at this stage since the full analysis with treatment interaction is best performed using the **fit** command in Genstat. We can, however, reflect on what the full blocking structure would be, namely:

$$\text{block} \qquad \text{repl / (mainpl * row)}$$

leading to the strata *repl*, *repl.mainpl*, *repl.row* and *repl.mainpl.row*. The final stratum is synonymous with *repl.mainpl.subpl* from the split-plot analysis.

5

Analysis across sites

5.1 Introduction

So far we have looked at the setting up and analysis of individual field trials. This has involved concepts of data collection and screening (Chapter 3) and the analysis of variance of summary data (Chapter 4). Chapters 7 and 8 will introduce more complex experimental designs and analyses which can lead to more precise estimation of treatment means. In a thorough program of treatment comparisons there would be a number of trials, each providing estimated treatment means. Each trial might require a different experimental design structure in order to address site dimensions and variation. Hence, regardless of the design used for an individual trial, it is usually better to carry out individual site analyses resulting in a set of estimated treatment means and a standard error for comparison. Thus with the overall task of evaluating treatments we are at the stage of having obtained unbiased estimates for treatment means from each trial analysis. These means can be collected into a two-way table indexed by sites and treatments ready for further analysis (Patterson and Silvey 1980). Other summary information which should also be collected and retained from the individual analyses includes the residual and treatment mean squares.

The next thing is to analyse the two-way table to see if the several sets of estimated treatment means can be represented by just one set of means. In other words, we need an analysis of variance with two factors (*site* and *treatment*) to see if the interaction is significant. If there is no site-by-treatment interaction, the two-way table can be represented by its margins (the factor main effects) and we would have a very simple summary of our series of trials in terms of the relative performance of the treatments. Unfortunately it is not usually as simple as that as the interaction (commonly known as genotype-by-environment interaction) is nearly always statistically significant, implying that the relative performance of treatments is changed from site to site.

In this *Eucalyptus grandis* trial at Mtao, Zimbabwe, an old bullock track runs through one replicate. Repeated ploughing during site preparation was not sufficient to remove the track's effects several years later. Smaller trees in the centre of the photo are on the track, larger trees to the left and right are off it.

Most of this chapter is devoted to a simple method for interpreting genotype-by-environment interaction. The method, called 'joint regression analysis', is only one of the large number of techniques which may elucidate structure in a two-way table when interaction is known to be present. Other popular methods and references to earlier work are discussed by Williams *et al.* (1992). When the two-way table is complete the joint regression analysis technique is more commonly known as Yates-Cochran (1938) or Finlay-Wilkinson (1963) analysis. Frequently, some treatments are not represented at all sites and the table will be incomplete; then the method of Yates and Cochran has to be generalised as shown by Digby (1979).

In Section 5.2 we introduce a data set from a series of field trials in Thailand and show how our philosophy of carrying out the analysis in stages is pursued across sites (c.f. the analysis of plot means in Section 3.2.3). An alternative approach would be to take plot summary data or even individual measurement data from several sites and carry out an overall analysis. This approach facilitates the calculation of breeding values and other genetic quantities (Chapter 6) but can have disadvantages, some of which are discussed in Section 5.3. On the other hand experimenters often choose to distribute limited treatment resources across multiple sites. Then the use of partially replicated designs (Section 7.7) with treatments specified as random effects (Chapter 6) can be used.

5.2   Complete two-way tables

We might have experiments at l locations or sites, at each of which v treatments are compared, and wish to combine the information from each site into one analysis of variance. If each individual experiment is an RCB design with r replicates the analysis of variance table at each site would have the following form for the jth site:

Source of variation	d.f.	m.s.
repl stratum	$r-1$	
repl.plot stratum		
treatment	$v-1$	
Residual	$(r-1)(v-1)$	s_j^2

We assume that the analysis is based on the plot means from each experiment. An extra stratum would be needed to carry out the analysis at the *trees* level but, as discussed in Chapter 3, this is not recommended. The estimated treatment means are obtained from each individual analysis. These are then assembled across the l sites, yielding a $v \times l$ table of estimated treatment means which can be analysed as a two-way analysis of variance with the form:

Source of variation	d.f.	m.s.
site	$I - 1$	
treatment	$v - 1$	
site.treatment	$(I - 1)(v - 1)$	

Is the site-by-treatment interaction (*site.treatment* in the above table) significant? The error term (S^2) for testing the interaction is the average of the s_j^2 from the individual analyses divided by the number of replications at each site. Before taking the average, check that the s_j^2 are homogeneous and can be combined together. Patterson and Silvey (1980) suggest that a 10-fold range of residual mean squares can be tolerated when forming a pooled residual over sites. If the range is greater, a weighted analysis of the two-way table should be considered, where the weight for the estimated treatment means from the jth site is $1/s_j$. We normally do not have to resort to this complicated procedure.

Division by the number of replications at each site is necessary because in the two-way table we are dealing with estimated treatment means from r plots in each experiment, whereas the s_j^2 relate to the means from each plot. A similar operation was carried out in Chapter 3 where we scaled by the number of trees per plot when switching from tree data to plot mean data. Normally, in a structured multi-site series of trials the numbers of replications will be the same at each site. If this is not the case, a weighted analysis of the two-way table should be performed, where the weights are the number of replications at each site. The form of the analysis of variance table needed to construct the appropriate F-test for testing the site-by-treatment interaction is:

Source of variation	d.f.	m.s.
site	$I - 1$	
treatment	$v - 1$	
site.treatment	$(I - 1)(v - 1)$	
Residual	$I(r - 1)(v - 1)$	$S^2 = \sum s_j^2 / (Ir)$

Example 5.1

In 1985 species/provenance trials were laid out at six sites in Thailand as part of an ACIAR project extending over several years to investigate Australian multi-purpose tree species. The experimental design in each case was an RCB design with three replicates and the number of seedlots ranged from 30 to 42. Plots consisted of 5 × 5 trees with a 2 × 2 m spacing. Plot summary files were constructed for the 24-month measurement according to the methods described in Chapter 3. Analyses were performed on the plot mean data at each site. Further

Table 5.1 Estimated means for the 24-month measurement of height (cm) at six sites in Thailand. (Reproduced from Table 1 of Williams and Luangviriyasaeng 1989)

Seedlot number	Species	Ratchaburi	Sai Thong	Si Sa Ket	Sakaerat	Chanthaburi	Huai Bong
13877	ACAAUL	334	503	318	304	163	124
13866	ACAAUL	348	413	399	251	205	196
13689	ACAAUL	424	742	606	463	288	254
13688	ACAAUL	439	802	570	375	363	200
13861	ACAAUR	465	858	661	523	379	345
13854	ACAAUR	428	899	658	527	391	316
13686	ACAAUR	567	884	–	–	–	–
13684	ACAAUR	520	941	641	538	363	293
13684	ACACIN	326	571	353	440	166	–
13863	ACACRA	416	799	677	453	–	–
13683	ACACRA	600	1083	738	658	285	–
13681	ACACRA	571	920	679	610	–	272
13680	ACACRA	584	1073	674	–	248	–
14623	ACADIF	608	695	–	–	–	–
14175	ACAFLA	436	624	421	271	137	–
14660	ACAHOL	552	685	415	433	336	251
13691	ACALEP	553	853	735	–	248	167
13653	ACALEP	520	796	658	392	306	189
13846	ACAMAN	473	483	521	283	143	–
13621	ACAMAN	419	497	453	166	187	102
14176	ACAMEL	–	259	158	161	174	134
13871	ACAPOL	256	392	261	177	141	96
14622	ACASHI	467	–	392	205	–	–
13876	ALLLIT	–	444	311	226	84	136
13519	CASCUN	500	434	420	334	186	176
13514	CASCUN	523	349	330	273	143	123
13148	CASCUN	383	267	220	223	132	158
13990	CASEQU	406	263	252	206	171	–
14537	EUCAM	764	884	839	632	569	337
14106	EUCAM	764	939	809	575	555	397
12013	EUCPEL	673	894	566	332	382	327
14130	EUCTOR	490	475	414	310	415	237
14485	MELBRA	216	116	97	63	80	75
14166	MELDEA	209	–	338	174	–	119
11935	MELDEA	218	319	363	191	76	100
14170	MELSYM	248	–	353	235	–	142
14152	MELVIR	–	–	288	90	111	113

details are given by Williams and Luangviriyasaeng (1989). Table 5.1 reproduces, from chapter 14 of the monograph, the estimated seedlot means for tree height. As can be seen from gaps in this table, not all the species were grown at each site. In addition, a number of local seedlots were incorporated in the trials; these seedlots were specific to each site and provided an important comparison with the Australian species but have been excluded from Table 5.1. It is common for site-by-seedlot tables to be incomplete, but for the moment we want to demonstrate the analysis of variance for a complete table so we will drop out the final two sites and the seedlots which are not present at all of the first four sites, to give the reduced file of means in Table B5.1. This file is in the form recommended in Section 3.2. A Genstat program to read the data in Table B5.1 and produce a two-way analysis of variance is given in Table C5.1. Note that in this analysis there is no term for the residual mean square. As already discussed, we obtain this quantity (S^2) from the pooled residual from each individual analysis and divide by $r = 3$ because the data in Table 5.1 are means of three replicates, whereas the pooled residual mean square is from data based on plot means. A similar scaling was carried out in Section 3.2.3 when combining results at the *trees* level with those at the *plots* level.. In Example 5.2 we will look in more detail at the calculation of S^2 but for now we will use a calculated value of $S^2 = 1040$ on 208 d.f. So the analysis of variance in Table 5.2 can be manually modified to give Table 5.3, which is now a completed example of the analysis of variance table presented earlier in this section.

The F-ratio in Table 5.3 is highly significant when the site-by-seedlot interaction mean square is tested against S^2. We will investigate this interaction further to see if we can understand what is causing the interaction and, hopefully, interpret the result; this is done using the method of joint regression analysis in Section 5.4. Before that, however, we will demonstrate the analysis of incomplete two-way tables, such as Table 5.1, using the **fit** command in Genstat.

Table 5.2 Genstat output from analysis of variance of estimated height means for Example 5.1.

Analysis of variance

Variate: v[1]; ht – mean height in cm

Source of variation	d.f.	s.s.	m.s.	v.r.	F pr.
site	3	919585.	306528.		
seedlot	26	3176289.	122165.		
site.seedlot	78	707957.	9076.		
Total	107	4803831.			

Tables of means

Variate: v[1]; ht – mean height in cm

Grand mean 488.5

site	Ratchaburi	Sai Thong	Si Sa Ket	Sakaerat
	461.6	627.7	494.0	370.5

seedlot	13877	13866	13689	13688	13861	13854	13684
	364.8	352.8	558.8	546.5	646.8	628.0	660.0

seedlot	13864	13863	13683	13681	14175	14660	13653
	422.5	586.2	769.8	695.0	438.0	521.2	591.5
seedlot	13846	13621	13871	13519	13514	13148	13990
	440.0	383.8	271.5	422.0	368.8	273.3	281.8
seedlot	14537	14106	12013	14130	14485	11935	
	779.8	771.8	616.2	422.3	123.0	272.8	

Table 5.3   Analysis of variance table for Example 5.1 with pooled residual mean square added manually.

Analysis of variance

Source of variation	d.f.	s.s.	m.s.	v.r.	F pr.
site	3	919585.	306528.		
seedlot	26	3176289.	122165.		
site.seedlot	78	707957.	9076.	8.73	<.001
Residual	208		1040.		

Acacia peregrinalis (seedlot no. 13689) and *A. holosericea* (seedlot no. 14660) at Ratchaburi, Thailand. This is one of the species/provenance trials discussed in Example 5.1. (Photo Khongsak Pinyopusarerk)

5.3   Incomplete two-way tables

A two-way table with gaps in it is known as an incomplete or non-orthogonal two-way table. There are complications in trying to analyse an incomplete two-way table. The estimated means are no longer derived as simple row or column averages. The reason for this is fairly clear. Suppose a particular treatment was included at all sites except the best site. It would not be fair to estimate that treatment mean by the average result over all sites for which it was present. The treatments that were included at the best site would have an advantage using this scheme. It is better to propose a statistical model for the two-way table then estimate the main effects using least squares. The linear model assumes that an entry or data item in the two-way table can be broken into three components: first, an overall mean; second, an adjustment to the mean to account for the particular treatment; third, another adjustment for the particular site. We write this as:

$$(\text{observation}) = (\text{overall mean}) + (\text{site effect}) +$$
$$(\text{treatment effect}) + (\text{site.treatment effect})$$

or symbolically:

$$Y_{ij} = \mu + \theta_i + \tau_j + \theta.\tau_{ij}$$

(5.1)

where the Y_{ij} ($i = 1,2,...,l$; $j = 1,2,...,v$) are entries in a site-by-treatment table; μ is a parameter for the overall mean; the θ_i and τ_j are parameters for the sites and treatments respectively; and the $\theta.\tau_{ij}$ are the interaction effects between sites and treatments. A complication in fitting this model to incomplete tables is that the final analysis of variance table can change depending on whether sites or treatments are fitted first. This is because the first term fitted is not adjusted for the second. It is normal to fit sites first then adjust the sums of squares for treatments for sites, thereby obtaining a more accurate indication of the overall variation between treatments. In Genstat the factors are fitted in the order that they are specified.

   Incomplete two-way tables can be analysed in Genstat using the **fit** command. The **anova** directive in Genstat can also be used for incomplete tables by replacing the gaps in the table with estimated missing values, but the resulting analysis of variance is only approximate and becomes more so as the number of gaps increases. It would be unwise to use **anova** if more than two or three missing values have to be estimated.

Example 5.2
A Genstat program to carry out the two-way analysis of Table 5.1 is presented in Table C5.2. The program assumes that data are in recommended form, as illustrated in Table B5.1. A difference here is that although the data file is of length 37 × 6 = 222, 30 of the height means are missing as indicated with the

symbol * in Genstat (Appendix A). The resulting analysis of variance table is given in Table 5.4. Just as with Table 5.3, the pooled residual mean square from the analyses of individual experiments can be appended to Table 5.4. Details of residual mean squares from the analyses of height means at each of the six sites are included in Table 5.5. This table gives an overall summary of results from individual analyses of variance, as well as the pooled and scaled plot variances obtained from the plot summary file as discussed in Section 3.2.3. Table 5.5 helps to determine whether homogeneity of variance assumptions are satisfied. The range of residual mean squares (s_j^2) is from 891 to 5852 cm^2. As mentioned in the previous section, a 10-fold range of residual mean squares is tolerated before a weighted analysis is deemed necessary. Such a range is greater than what would be dictated by a ratio test of variances, but we prefer to follow the pragmatic view of Patterson and Silvey (1980). Thus the unweighted analysis in Table 5.4 is supported by the results in Table 5.5. The next thing is to form the pooled and scaled residual mean square (S^2), which can be appended to Table 5.4. There is no need to be too rigid about the way this is done. Strictly speaking, the pooled residual should be calculated as a weighted average of residual mean squares of individual experiments where the weights are the corresponding degrees of freedom. But in most variety trial programs where seedlots are being compared across a range of sites, the residual mean squares all have roughly the same degrees of freedom so it is easier to simply take the average of the residual mean squares. We would have to be a little more careful if the numbers of degrees of freedom for the residual mean squares were small, e.g. less than 20. In this example there are adequate degrees of freedom for the residual mean square at each site so it is acceptable to take just the average of the residual mean squares in Table 5.5. This value is 2891 cm^2, which is then divided by the number of

---

Table 5.4   Genstat output from analysis of variance of estimated height means for Example 5.2.

Regression analysis

> Response variate: v[1]; ht – mean height in cm
> Fitted terms: Constant + site + seedlot + site.seedlot

Accumulated analysis of variance

Change	d.f.	s.s.	m.s.
+ site	5	4157543.	831509.
+ seedlot	36	4425296.	122925.
+ site.seedlot	150	1351054.	9007.
Total	191	9933893.	52010.

Table 5.5 Summary of analysis of variance mean squares for estimated height means from individual sites.

Site	Replicate	Seedlot	Mean squares Residual	Within plot
Ratchaburi	34376	63700	3068	364
Sai Thong	85252	216149	5852	846
Si Sa Ket	13521	112267	1866	430
Sakaeret	61971	82155	1804	381
Chanthaburi	128691	58430	3865	526
Huai Bong	256	23893	891	274

Table 5.6 Analysis of variance table for Example 5.2 with pooled residual mean square added manually.

Accumulated analysis of variance

Change	d.f.	s.s.	m.s.	v.r.	F pr.
+ site	5	4157543.	831509.		
+ seedlot	36	4425296.	122925.		
+ site.seedlot	150	1351054.	9007.	9.35	<.001
Residual	384		964.		

replications at each site (i.e. three) to give $S^2 = 964$ cm^2. The degrees of freedom (384) for S^2 are the sum of residual degrees of freedom for each site. The summary analysis of variance table incorporating the pooled residual mean square from Table 5.5 into Table 5.4 is presented in Table 5.6. The F-test of the site-by-seedlot interaction on 150 and 384 d.f. is 9.35, which is very highly significant; this interaction will be investigated further in Section 5.5.

There is obviously a need to be careful when analysing data in stages then combining the results to get an overall analysis of variance table. We could easily forget to rescale the pooled residual mean square in Table 5.6. There may also be some approximations when, say the numbers of trees per plot or replication numbers of treatments are not equal. Nevertheless there are considerable advantages in a step-by-step approach to analysis compared with an overall analysis across all sites. Some points to consider are:

1. Analysing experiments separately in order to obtain estimated treatment means allows specialised software to be used. It is common to use incomplete block designs for the field trials (see Chapters 7 and 8). It is then necessary to use an analysis appropriate to the particular experimental design. Once the estimated treatment means are obtained, details of individual designs can to some extent be forgotten and we can proceed with our standard approach to the two-way table of treatment means over sites.

2. There is no need for the experimental designs at each site to be the same. Sometimes one site requires a type of plot configuration or blocking structure that is different from that required at another site. By analysing each site separately, it is easy to cater for different experimental layouts. All that are carried forward to the across-sites analysis are the estimated treatment means and details of the residual mean squares.
3. A single analysis possibly at the individual measurement level and incorporating the need for different blocking structures at each sites would automatically form a pooled residual mean square. Thus there would be no opportunity to obtain the individual residual mean squares from each site for checking that it is appropriate to form a pooled residual mean square.

5.4   Joint regression analysis with complete two-way tables

The analysis of variance table for Example 5.1 (Table 5.3) is fairly typical of the type of conclusion we obtain from the combination of seedlot means across sites; the site-by-seedlot interaction is usually highly significant. We are therefore interested in trying to explain the interaction. It is usually easy to see what is causing a significant interaction if the table of interactions is small. Suppose, for example, we have a 2×2 factorial experiment to assess the effect of adding nitrogen and phosphorus on the growth of trees. The levels of the factors N and P correspond to no added fertiliser (level l) and added fertiliser (level 2). From the analysis of variance the estimated means for the variate ht are as shown below.

	Nitrogen	
Phosphorus	1	2
1	2.3	4.5
2	3.6	10.2

It is clear that the combined effect of adding nitrogen and phosphorus is much greater than a prediction based on looking at the separate effects of adding nitrogen or phosphorus. Thus a significant interaction can easily be interpreted. With larger tables it is much more difficult to recognise the cause of interaction and statistical analysis is needed to identify and interpret a significant interaction. Interpretation is often aided by extra information such as the latitude and altitude of the seedlot sources.

A number of techniques are available for elucidating structure in two-way tables. The general terminology for such methodology is the analysis of genotype-by-environment interaction. In the case of Example 5.1, the genotypes and environments correspond to the factors *seedlot* and *site* respectively. We will concentrate on the method of joint regression analysis to seek the cause of site-by-treatment interaction. This method is probably the simplest technique available and tries to characterise the interaction by a set of regression coefficients, one for each treatment. The model is:

(observation) = (overall mean) +

(treatment regression coefficient) × (site effect) +

(treatment effect) + (residual)

or symbolically:

$$Y_{ij} = \mu + \gamma_j \, \theta_i + \tau_j + \varepsilon_{ij} \tag{5.2}$$

This model contains extra parameters not present in the simple model (5.1). These parameters are the treatment regression coefficients (γ_j), introduced as multipliers for the site effects. If there is no interaction, all the treatment regression coefficients are equal to one and the model simplifies to our earlier version. Normally there will be some interaction and the γ_j will vary about one; some will be higher and some will be lower.

Estimation of the parameters in (5.2) is not difficult for complete two-way tables and will be outlined in this section. Apart from the overall mean, three sets of parameters are to be estimated, namely the site effects, treatment effects and the treatment regression coefficients. For complete two-way tables, the site and treatment effects are obtained from the margins of the table and hence correspond to the estimators in Section 5.2. Once we have the estimated site effects, the treatment means for each site are regressed on them to give a regression coefficient for each treatment. More commonly, the estimated site means are used in the regression rather than estimated site effects, but this is just a matter of adding the overall mean to each effect. Further description of the technique is best done using an example, so the subset of data from Thailand, analysed in Example 5.1, will be investigated further.

Example 5.1 (continued)
The estimated site and seedlot means for the Table B5.1 data are given in Table 5.2. The estimated regression coefficients can be calculated by carrying out the regression of the individual seedlot means over sites on the overall site means. The Genstat syntax for obtaining these regression coefficients is given later in an overall program (Table C5.3). From Table 5.2 the estimated site means are 462, 628, 494 and 371 cm. Now, in Table 5.1 look at seedlot number 13653 (*Acacia leptocarpa*) which has site values of 520, 796, 658 and 392 cm respectively for the first four sites. The linear regression of *Acacia leptocarpa*

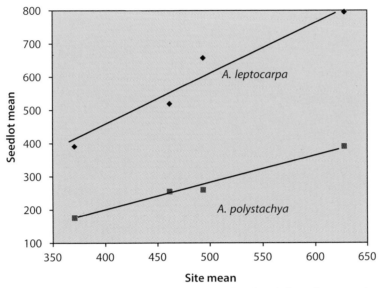

Fig. 5.1. Plot of mean height of two seedlots versus mean height of all seedlots at each of four sites.

values on the overall site means has a regression coefficient of 1.59. On the other hand, seedlot number 13871 (*Acacia polystachya*), with site values of 256, 392, 261 and 177 cm, has a regression coefficient of 0.83. The situation for these two seedlots is depicted in Fig. 5.1, where the site means are plotted on the horizontal axis and have been ordered so that the worst site (i.e. lowest estimated site mean) comes first and the best site last. The regression lines and individual points are shown in the figure. With *Acacia leptocarpa*, the slope of the regression line is greater than one; this means that the species is showing a big difference in performance between sites, growing poorly at poor sites and well at good sites. Such a species is called 'unstable'. But with *Acacia polystachya* the regression coefficient is less than one, which means that the species tends to perform more consistently than average over both poor and good sites. This type of seedlot is said to be 'stable'.

In terms of selecting suitable species based on the results of series of field trials, knowledge about the stability of species over sites can be important. Provided they have similar overall growth rates, a stable species might be preferred for planting at a range of sites of variable quality, whereas an unstable species would be better if the intended sites were known to be good. Table 5.7 contains the calculated regression coefficients for all 27 seedlots; for convenience we have repeated the estimated height means from Table 5.2.

Table 5.7 Estimated height means and regression coefficients for data from Table 5.2.

Seedlot number	Mean	Regression coefficient
13877	365	0.79
13866	353	0.61
13688	559	1.20
13689	547	1.73
13861	627	1.44
13854	628	1.62
13684	660	1.68
13864	423	0.61
13863	586	1.48
13683	770	1.80
13681	695	1.31
14175	438	1.34
14660	521	0.93
13653	592	1.59
13846	440	0.71
13621	384	1.20
13871	272	0.83
13519	422	0.29
13514	369	0.12
13148	273	0.05
13990	282	0.08
14537	780	0.96
14106	772	1.38
12013	616	2.07
14130	422	0.55
14485	123	0.10
11935	273	0.53

The joint regression analysis model requires addition of seedlot regression coefficients to identify the cause of site-by-seedlot interaction. We have already discussed how the regression coefficients can be interpreted in terms of stable and unstable seedlots, but we have not assessed how good the joint regression model is. The regression coefficients are of little value if they do not help to account for the interaction. We can partition the interaction sums of squares in Table 5.3 into two components: (i) the sum of the regression sums of squares over seedlots and (ii) deviations from regressions. If the regression mean square is significant and the mean square for the deviations is non-significant, the joint regression analysis model has been successful in explaining the interaction. The joint regression analysis of variance table for our example is given in Table 5.8. The mean square for regressions is 11866, compared with only 7682 for the deviations. Clearly the deviations are still highly significant, which means that the joint regression analysis model is not completely successful in summarising the interaction. Nevertheless the seedlot regression coefficients in Table 5.7 can still provide an explanation for some of the interaction. Further discussion of this point is given by Williams *et al.* (1992). It is very useful to plot the seedlot regression coefficients against seedlot means, as in Fig. 5.2. The unstable seedlots with high means are in the upper right part of the figure; these are often considered to be seedlots of interest although, as mentioned earlier, stable seedlots with high means (lower right of figure) can also be important. A Genstat program to produce all the results of this section, including the plot of seedlot regression coefficients against seedlot means, is given in Table C5.3.

Table 5.8 Joint regression analysis of variance for a complete site-by-seedlot table with pooled residual mean square added manually.

Analysis of variance

Source of variation	d.f.	s.s.	m.s.	v.r.	F pr.
site	3	919585.	306528.		
seedlot	26	3176289.	122165.		
site.seedlot	78	707957.	9076.		
regressions	26	308503	11866		
deviations	52	399454	7682	7.39	<.001
Residual	208		1040.		

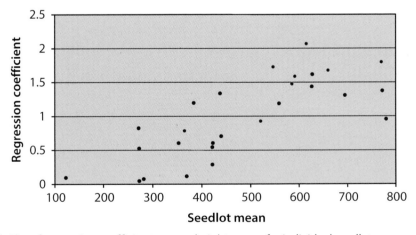

Fig. 5.2. Plot of regression coefficients versus height means for individual seedlots.

Provenance trial of *Acacia crassicarpa* (aged 18 months) in Thailand with Doug Boland. This is part of the 1986 ACIAR series of trials. (Photo Khongsak Pinyopusarerk)

5.5 Joint regression analysis with incomplete two-way tables

More often than not with a series of experiments, the resulting site-by-treatment table of estimated means is incomplete. Some reasons for this are, for example in field trials:

1. Seedlots have not been included at every site because it is known in advance that the seedlots will perform badly, e.g. a frost-susceptible seedlot at a high-altitude site.
2. Seedlots have been included at every site but have not survived at every site.
3. Some seedlots have insufficient seedlings to be planted at every site.

It is more difficult to apply the joint regression analysis model to incomplete two-way tables. The model remains the same as in the previous section; the problem comes with the estimation of the three sets of parameters. No longer can we simply estimate the site and treatment means by table margins, as can be done with a complete table, leaving only the estimated regression coefficients as the tricky part. Instead, all parameters for incomplete two-way tables require some computational effort. The problem can be overcome by excluding the treatments which are not represented at all sites. Although this is in fact how Table B5.1 has been derived from Table 5.1, it is not a good solution since an important treatment that is present at all but one site cannot be ignored. On the other hand, a treatment that is represented at only one site in the two-way table is quite likely to be a poorly performing species that has not survived at other sites. It may be better to exclude this treatment from the joint regression analysis as treatments with poor site representation in the two-way table may cause computational difficulties and add little information to the analysis. A common compromise is to ensure that treatments are represented at no fewer than three sites. If treatments at fewer sites are included in the analysis, we must make very sure that the estimated treatment means and regression coefficients are sensible. This is an important consideration when choosing parameters for the construction of partially replicated designs across multiple sites (Section 7.7).

Ultimately we need a method for estimating the joint regression analysis parameters for an incomplete two-way table. Such a technique was suggested by Digby (1979), involving an iterative approach to the parameter estimation. The idea is to guess one set of parameters, then estimate the other two sets and hence obtain a new guess for one of the sets, estimate the other two and so on until the three sets of estimated parameters converge to a stable result. We will not go into details of the method, as the main thing to understand is that the method yields estimated parameters that have the same interpretation as in the previous section. The extra complications introduced by an incomplete two-way table means that more care is required in constructing an analysis of variance table analogous to Table 5.8. We need to define the problem in terms of non-linear analysis as explained by Ng and Williams (2001) and illustrated in the following example.

Example 5.2 (continued)

A Genstat program to carry out joint regression analysis on the data in Table 5.1 is given in Table C5.4. The program assumes that the data are in the recommended form (illustrated in Table B5.1) and that the data file is ordered by site and seedlot. Results from the program are given in Table 5.9. Note that we have included in the analysis the seedlots represented at only two sites in Table 5.1, namely seedlots 13686 and 14623. Because the method can have difficulty in separating seedlot means and regression coefficients for poorly represented seedlots, it is necessary to check that spurious results are not obtained. In fact the plot of regression coefficients against seedlot means (Fig. 5.3) shows that seedlot 14623 (No. 14 in the figure) is somewhat isolated and it would probably be appropriate to re-run the analysis excluding this seedlot. The numbers in Fig. 5.3 correspond to those in the first column of Table 5.9 to allow easy identification of the seedlots. The figure highlights the fact that seedlots Nos. 11 and 13 (both *Acacia crassicarpa*) are unstable. The tallest seedlots are Nos. 29 and 30 (both *Eucalyptus camaldulensis*). It is evident from Fig. 5.3 that provenances from the same species are tending to group together. We highlight this by re-doing the figure using the species codes (a–v) in Table 5.9 (see Fig. 5.4). It shows very effectively how results of the trials

Table 5.9 Estimated height means and regression coefficients for data from Table 5.1.

(a) Seedlot

No.	Seedlot number	Code	Species	Mean	Regression coefficient
1	13877	a	*Acacia aulacocarpa*	291	0.86
2	13866	a	*Acacia aulacocarpa*	302	0.60
3	13689	a	*Acacia aulacocarpa*	463	1.18
4	13688	a	*Acacia aulacocarpa*	458	1.30
5	13861	b	*Acacia auriculiformis*	539	1.20
6	13854	b	*Acacia auriculiformis*	537	1.29
7	13686	b	*Acacia auriculiformis*	510	1.63
8	13684	b	*Acacia auriculiformis*	549	1.46
9	13864	c	*Acacia cincinnata*	336	0.92
10	13863	d	*Acacia crassicarpa*	465	1.49
11	13683	d	*Acacia crassicarpa*	594	2.04
12	13681	d	*Acacia crassicarpa*	572	1.46
13	13680	d	*Acacia crassicarpa*	519	2.25
14	14623	e	*Acacia difficilis*	592	0.45
15	14175	f	*Acacia flavescens*	327	1.33
16	14660	g	*Acacia holosericea*	445	0.92
17	13691	h	*Acacia leptocarpa*	500	1.73
18	13653	h	*Acacia leptocarpa*	477	1.46
19	13846	i	*Acacia mangium*	342	1.01
20	13621	i	*Acacia mangium*	304	1.02
21	14176	j	*Acacia melanoxylon*	179	0.23
22	13871	k	*Acacia polystachya*	220	0.68
23	14622	l	*Acacia shirleyi*	304	1.62
24	13876	m	*Allocasuarina littoralis*	246	0.81
25	13519	n	*Casuarina cunninghamiana*	342	0.73
26	13514	n	*Casuarina cunninghamiana*	290	0.66

27	13148	n	*Casuarina cunninghamiana*	230	0.34
28	13990	o	*Casuarina equisetifolia*	250	0.25
29	14537	p	*Eucalyptus camaldulensis*	671	1.24
30	14106	p	*Eucalyptus camaldulensis*	673	1.26
31	12013	q	*Eucalyptus pellita*	529	1.33
32	14130	r	*Eucalyptus torelliana*	390	0.46
33	14485	s	*Melaleuca bracteata*	108	0.14
34	14166	t	*Melaleuca dealbata*	226	0.67
35	11935	t	*Melaleuca dealbata*	211	0.67
36	14170	u	*Melaleuca symphyocarpa*	261	0.67
37	14152	v	*Melaleuca viridiflora*	189	0.58

(b) Sites

No.	Site	Mean
1	Ratchaburi	425
2	Sai Thong	620
3	Si Sa Ket	481
4	Sakaerat	359
5	Chanthaburi	258
6	Huai Bong	198

across sites can be summarised at a species level. The four provenances of *Acacia crassicarpa* (code d) are grouped in a desirable position on the figure (high seedlot means and regression coefficients). This is also the case with the four provenances of *Acacia auriculiformis* (code b). Of particular interest is the fact that the four provenances of *Acacia aulacocarpa* (code a) have split into two groups of two. Further investigation shows that the seedlots of one group are from Papua New Guinea and the seedlots from the other group are from Queensland in Australia. The Papua New Guinea seedlots (Nos. 3 and 4) have grown taller than the Queensland seedlots (Nos. 1 and 2) and demonstrate clear variation between groups of provenances of the same species. In fact, subsequent to these trials, a taxonomic revision of *Acacia aulacocarpa* (McDonald and Maslin 2000) recognised the Papua New Guinea populations as a separate species, *A. peregrinalis*, and identified several different species within the Australian populations. Figures 5.3 and 5.4 represent a very convenient summary of the complete data set. But, as we pointed out in the previous section, there must be some check that the joint regression analysis is accounting for enough interaction to be retained.

The analysis of variance table (obtained using the non-linear analysis method of Ng and Williams 2001) is given in Table 5.10. We can compare the joint

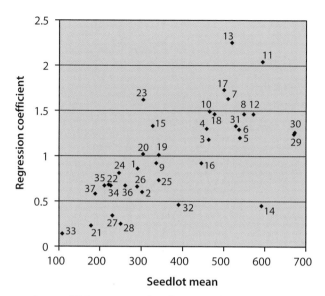

Fig. 5.3. Plot of regression coefficients versus height means for individual seedlots.

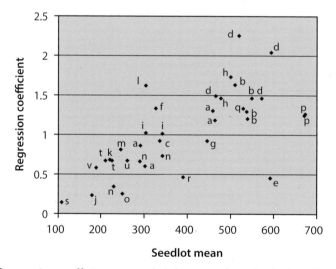

Fig. 5.4. Plot of regression coefficients versus height means for individual seedlots, with each species coded.

regression model against a model which fits only main effects, by looking at the ratio of the regressions and deviations mean squares. This gives an F-test of 4.13 on 36 and 114 d.f. which is highly significant (P < 0.001), so the impact of the joint regression model is substantial. The F-test of deviations against the pooled residual

Table 5.10 Joint regression analysis of variance for an incomplete site-by-seedlot table with pooled residual mean square added manually.

Accumulated analysis of variance

Change	d.f.	s.s.	m.s.	v.r.	F pr.
+site	5	4157543.	831509.		
+seedlot	36	4425296.	122925.		
+site.seedlot	150	1351054.	9007.		
regressions	36	764664.	21241.	4.13	<.001
deviations	114	586390.	5144.	5.34	<.001
Residual	384		964.		

mean square is, however, also still highly significant and so, even though Figs 5.3 and 5.4 provide a very valuable interpretation of the interaction, we must be aware that there may still be the potential to interpret further information from the interaction. The Table 5.10 analysis of variance can be pieced together from Table 5.6 and the residual sum of squares from the output from the program in Table C5.4. But as Ng and Williams (2001) point out, the degrees of freedom for this residual must be calculated by difference from Table 5.10, rather than using the value from the Table C5.4 output.

Finally, note that for this example the nested treatment structure (Section 8.4) of species and provenances within species could have been incorporated into the analysis of variance in Table 5.6. We did not do this because we wanted to demonstrate the sequence of events for the simplest two-way table, namely sites by seedlots. Partitioning the seedlots in such a way would generate an interaction table for both sites by species and sites by provenances within species. Depending on the relative significance of these interaction terms in the analysis of variance table, we could choose to apply the joint regression analysis to the estimated site-by-species means. The results in Fig. 5.4 suggest that, provided we partition the *Acacia aulacocarpa* provenances into two separate species, a site-by-species joint regression analysis could provide an effective summary of the data.

6
Variance components and genetics concepts

6.1 Introduction

All linear models considered in Chapter 4 involved estimation of means for the levels of each factor and a residual mean square for testing differences between estimated means. In this chapter we introduce a different kind of model where, for some factors, we want to estimate single quantities called 'variance components' instead of estimating means. We will introduce the method of residual maximum likelihood (REML) to estimate variance components and demonstrate the use of REML in Genstat.

We will give an overview of a number of genetics concepts, such as genetic variance, heritability and genetic correlation. Variance components are fundamental to these concepts. In Sections 6.3–6.6 we demonstrate how the variance components estimated from our analysis can be used to calculate these quantities in the common situation of trials testing sets of open-pollinated families. In Section 8.5 we demonstrate an analogous approach for the analysis of clone trials.

6.2 Variance components

Models to explain sources of variation are often written in terms of variance components. Examples include:

1. Multiple-tree plot experiments where we want to separate the variation between trees within a plot from that between plots, as discussed in Chapter 3.

2. Progeny and provenance/progeny trials and clone trials where variance components are used to estimate genetic variation (Falconer and Mackay 1996, chapter 8).

3. Mixed-effects models for analysis of incomplete block designs, as will be discussed in Chapters 7 and 8.

4. Industrial experiments and quality control where variance components are used to isolate areas requiring improvement in performance.

In this book we will not go into details of variance component models and estimation. A general account of variance components in genetics is given by Falconer and Mackay (1996, chapter 8) and in statistics by Snedecor and Cochran (1989, chapter 13). We will look at only two very common situations where variance components arise in tree improvement trials, namely points 1 and 2, above.

6.2.1 Stratum variance components

In Section 2.5 we discussed how strata can be used to account for field variation. We try to account for this variation in the underlying linear model. Typically we assume a level of random variation for each stratum and that these random effects are independent. To be more specific, we will look at the example of a multiple-tree plot experiment. The bottom two strata in the analysis would be the between-trees-within-plots stratum (*trees* level) and the between-plots stratum (*plots* level). We assume that the variance between any two trees within a plot is $2\sigma_t^2$, where σ_t^2 is called the *trees* level variance component. The variance between two trees in different plots would be $2(\sigma_t^2 + \sigma_m^2)$, where σ_m^2 is the *plots* level variance component. As mentioned in Section 2.5, we often choose to interpret σ_t^2 as arising mainly from genetic variation between trees of a seedlot, although it actually includes non-genetic effects, such as nursery effects, propagation effects in clone trials and micro-environment differences within the plot. This is an example of the assumptions that we (perhaps unknowingly) make when postulating a linear model, carrying out an analysis and interpreting results.

The variance components σ_t^2 and σ_m^2 have to be estimated, and there are many ways of doing this. Consistent with the approach in Chapter 3 of analysing plot means, we recommend the following method:

1. The estimator $\hat{\sigma}_t^2$ be obtained by pooling the plot variances (Section 3.2.3). When there are missing trees, we might prefer to take a weighted average of the plot variances but, as discussed in Section 3.2.3, this complication is usually not necessary.

2. The residual mean square (s^2) from the analysis of variance of plot means provides a mechanism for obtaining $\hat{\sigma}_m^2$. From the specification of the linear

model we can derive the expected value of s^2, namely $\sigma_t^2/w + \sigma_m^2$, where w is the number of trees per plot. So, using $\hat{\sigma}_t^2$ from point 1 we obtain:

$$\hat{\sigma}_m^2 = s^2 - \frac{1}{w}\hat{\sigma}_t^2 \tag{6.1}$$

This method of estimation is known as 'equating mean squares to expectation' and is a convenient way of partitioning s^2 into its components. Usually there will be missing trees and so the number of trees per plot is likely to be different from plot to plot. Then w in (6.1) must be replaced by $\bar{w}$, the harmonic mean of the tree counts as discussed by Cochran and Cox (1957, section 14.31). Note that from a plot summary file it is possible to obtain $\bar{w}$ and $\hat{\sigma}_t^2$ easily.

Example 6.1

Take the data used in Example 3.1. From Table 3.4 the average of the plot variances gives $\hat{\sigma}_t^2 = 2.509$ and of course $w = 8$. From Table 3.5 we get $s^2 = 0.7475$ and hence $\hat{\sigma}_m^2 = 0.434$. As discussed in Section 3.2.3, there are strong grounds to exclude three data values, namely units 28, 51 and 76. Re-doing the plot summary file gives the estimate $\hat{\sigma}_t^2 = 0.766$ and the harmonic mean of seven plots with eight trees and three plots with seven trees is $\bar{w} = 7.671$. The (unweighted) analysis of plot means gives $s^2 = 0.252$ and so $\hat{\sigma}_m^2 = 0.152$. As expected, dropping out the three unusual data values has a very big effect on the estimated stratum variance components.

6.2.2 Treatment variance components

The linear models in Chapter 4 consist of parameters for factors of interest. These parameters are then estimated for each level of each factor. For example, the RCB model:

$$Y_{ij} = \mu + \rho_i + \tau_j + \varepsilon_{ij} \tag{6.2}$$

from Section 4.2.2 contains r parameters (ρ_i) for the blocking factor *repl* and v parameters (τ_j) for the treatment factor. This model is known as a fixed-effects model because the overall mean, replicate and treatment parameters are assumed to be fixed and requiring estimation, compared with the ε_{ij} which are assumed to be random effects with zero mean and variance σ^2. These fixed-factor estimates are called Best Linear Unbiased Estimates (BLUEs). It is possible in forestry experiments for seedlots to be specified as fixed effects or as random effects. An example of fixed seedlot effects arises in the early stages of a tree introduction program. Seedlots are collected from many natural provenances of a species and from existing breeding programs elsewhere. They are tested together in a field experiment with the objective of estimating and determining the best seed sources, from which broader collections may be made to assemble the base population for breeding.

There are cases, however, where seedlots form a random sample of a large population of such seedlots. We are then interested in the genetic variance, i.e. that due to seedlot differences in the large population as estimated by the sample of seedlots which we are able to test. The model (6.2) is then formulated as:

$$Y_{ij} = \mu + \rho_i + \varphi_j + \varepsilon_{ij} \qquad (6.3)$$

where the φ_j are now random treatment effects with zero mean and variance σ_f^2, independent of ε_{ij}. The variance component σ_f^2 is then the quantity that is estimated, i.e. just one parameter rather than $v - 1$ in the fixed-effects model. We can, however, still obtain estimates for each treatment level; these are called Best Linear Unbiased Predictions (BLUPs) (Robinson 1991). An example of random seedlot effects is a breeding program where parents are randomly chosen from a breeding population and open-pollinated seed is obtained from them, or the chosen trees are crossed together following a mating design. The resulting seedlots are raised and planted in a field experiment in which an important objective is the estimation of variance components. Seedlot variance components can have a genetic interpretation which permits estimation of genetic parameters, such as heritability and genetic correlations. These parameters are important parts of equations to predict genetic gains from breeding programs. A model which incorporates both fixed and random effects is called a mixed-effects model; e.g. in (6.3) we have fixed replicate and random seedlot effects. Note that in Chapter 8, we consider a mixed model where some of the blocking factors are treated as random effects; this is done to allow the recovery of inter-block information in the estimation of treatment effects.

Parameter estimation in mixed-effects models is complicated because we must obtain estimated means for fixed effects and estimated variance components for random effects. The **anova** and **fit** commands in Genstat will only analyse fixed-effects models, but there is another command called **reml** in Genstat that will do the required estimation. The acronym REML stands for 'residual maximum likelihood' and is the recommended method of parameter estimation for mixed models (Patterson and Thompson 1971). The following example illustrates the use of REML.

Example 6.2

A progeny trial of *Acacia mangium* was planted at Segaluid, Sabah, Malaysia, by the Sabah Forest Research Centre in 1994. The trial was designed to test 48 open-pollinated families collected from natural provenances in Papua New Guinea (PNG, 41 families) and Far North Queensland (five families) and two families of the land race that had developed in Sabah after introduction of *A. mangium* in the 1960s. Based on the results of many previous trials (Harwood and Williams 1992), it was expected that the Sabah and Queensland families would perform more poorly than those from PNG. The trial was set out as an RCB design with four replicates each containing 48 five-tree plots. Spacing was 3 × 3 m between trees, and an external perimeter row surrounded the trial. Diameter at breast height (*dbh*) and height (*ht*) measurements were taken in 1997, 36 months after planting.

We shall concentrate initially on *dbh*. The plot summary file (Table B6.1) shows the indexing information for *repl, family, prov* and the plot means, variances and plot counts for *dbh*. For simplicity, we will first examine the data only for the 41 families from PNG, and consider these as representing a single provenance, ignoring the fact that they derive from a number of different local provenances within PNG, which display very similar performance. A Genstat program to read the data and estimate the seedlot variance component σ_f^2 is given in Table C6.1. We make use of the **restrict** command in Genstat to exclude the families from Queensland and Sabah from the calculation. Also included are statements to calculate the harmonic mean of the tree counts and estimates of the *trees* and *plots* variance components (σ_t^2 and σ_m^2 respectively) and the estimated heritability (h^2) for *dbh*, which will be discussed in Section 6.4. The Genstat output is presented in Table 6.1, whence we get estimates of variance components:

$$\hat{\sigma}_t^2 = 3.922$$
$$\hat{\sigma}_m^2 = 0.276 \tag{6.4}$$
$$\hat{\sigma}_f^2 = 0.258$$

and:

$$\bar{w} = 4.403$$

6.2.3 Other options

Procedures described in the above two sections for estimating variance components are our recommendations for these common situations. Other circumstances, such as complex experimental designs or very large data sets, may require different approaches. Some comments on alternatives are given here:

1. Stratum variance components can be estimated from an analysis of variance at the *trees* level, such as is given in Table 3.3. The mean squares in that table can be equated to expectation, but it should be noted that in this situation the Genstat **anova** mean squares for seedlots and the *plots* level residual are only approximate when there are missing values. Exact mean squares are given by the **fit** command in Genstat.

2. The **reml** command in Genstat can also be used for the estimation of stratum variance components at the *trees* level. In general, however, for the reasons given in Chapter 3, it is much better first to create a plot summary file for a

Progeny trial of *Acacia peregrinalis* growing at Bengkoka, Sabah. This fastigate tree would be an outlier in any analysis of growth traits and should be excluded. There were, however, several trees of this type in the trial, all of the same family, suggesting the fastigate trait may be under genetic control and that the trait should not be ignored.

multiple-tree plot experiment and use the analysis of plot means to estimate stratum variance components.

3. Even if we estimate σ_t^2 from the plot summary file then analyse the plot means to estimate σ_m^2, there are some options. In Section 6.2.1 we recommend the unweighted analysis of plot means, but a weighted analysis could have been done using plot counts as weights. Cochran and Cox (1957, section 14.31) discuss weighted versus unweighted analysis.

4. In Section 6.2.2 we looked at a single treatment factor, i.e. *seedlot*. But often the seedlots consist of families from a number of different provenances and we are interested in the family-within-provenance variance component. Example 6.2 is such a situation although at present we have restricted attention to just one

provenance, namely PNG. In Section 6.5 we discuss the estimation of between-families-within-provenances variance components.

5. For very large multiple-tree plot experiments, single-tree plot experiments with missing values or those involving more complex blocking structures (such as those discussed in Chapters 7 and 8), it may be necessary first to carry out an analysis ignoring any treatment structure. From this analysis, estimated seedlot means can be saved in a file then re-analysed using the appropriate treatment structure. This two-stage process is quite common with fixed-effects models but does raise some difficulties if treatment variance components are required. Fortunately, however, improvements in computational procedures in major statistical packages have considerably increased our ability to estimate parameters in large and complex mixed-effects models.

6.3   Genetics concepts

Here we introduce some principles of genetics to show the link between the field trials we plant and progress in a breeding program. More complete coverage of quantitative genetics is given by Falconer and Mackay (1996), and its application to forestry by Wright (1976), Namkoong et al. (1988) and White et al. (2007).

Breeding programs usually begin with selections from a base or starting population. Selections are made and crossed, and their progeny tested in progeny tests. Selections are made in the progeny tests and the process continues. Progress depends on several things:

1. The variability in the trees from among which we have to select parents. This variability can be expressed as the phenotypic standard deviation, σ_p, which is the square root of the phenotypic variance, usually estimated as the sum $\sigma_p^2 = \sigma_f^2 + \sigma_m^2 + \sigma_t^2$.
2. How intensively selection is carried out. This can be expressed as the selection intensity, i, which is the difference between the mean of the selected individuals and the overall population mean, divided by σ_p.
3. The proportion of the phenotypic variance due to genetic variation (heritability, h^2).

In algebraic terms, the response to selection is:

$$R = i\sigma_p h^2 \tag{6.5}$$

We choose parents to cross based on what they look like (their phenotype), hoping that this reflects the genes they carry (their genotype). The closer these two are aligned, the higher is the heritability.

At the beginning of a breeding program, the amount of genetic variation is usually directly related to how the base population was formed. In some early eucalypt breeding programs, the base population was very small. Hence there was not much genetic variation and little progress in breeding could be made (Eldridge *et al.* 1993, chapter 1). So it is essential for a successful breeding program to begin with an adequate sample of the relevant species to form the base population. What do we mean by 'adequate'? For long-term breeding to support a major plantation program, a base population of 200 or more unrelated families, preferably from known superior provenances of a species, is typically recommended (Eldridge *et al.* 1993; White *et al.* 2007). We then carry out field trials to estimate the genetic variance available in our base population for selection, and to predict what progress we can make. Typically, superior trees are selected in these same field trials, for

Researcher Piare Lal inspects two fast-growing eucalypt clones with contrasting crown characteristics in a trial in Punjab State, India.

ongoing breeding. These selected trees may be vegetatively propagated, for testing in clone trials or for incorporation in clonal seed orchards, or seed or pollen may be collected from them for subsequent testing and breeding.

Our estimates of genetic variance in a breeding population are subject to sampling variation and to the adequacy of our experimental design. Sampling variation in this context arises from our choice (sampling) of individuals from the population for testing in trials – very seldom do we test the entire breeding population. The adequacy of experimental design also affects both our estimates of genetic parameters and the accuracy of selections we make in progeny trials or clone trials, and thus has a fundamental effect on progress in breeding.

6.4   Heritability

Heritability for a trait in a population is the additive genetic variance that enables genetic gain to be passed on to the next generation through the selections we make, and is expressed as a proportion of the phenotypic variance (σ_p^2). The phenotypic variance is usually estimated as the sum of the seedlot variance component (usually estimated from families – see Section 6.2.2), the *plots* level variance component (σ_m^2) and the *trees* level variance component (σ_t^2). Where the analysis has been carried out on plot means, the residual variance contains both the *plots* level and the *trees* level variance components and we must calculate them separately, as illustrated in Section 6.2.1. For single-tree plots, there is no independent estimate of σ_t^2 and so the phenotypic variance contains only the genetic variance and σ_m^2. There are some things we can do to make sure the estimate of heritability is as good as possible. The magnitude of the genetic variance component is determined by the material under trial (the set of seedlots that we choose to test) and the environment in which it is tested, which affects how genetic differences are expressed. An adequate sample, preferably more than 100 families (White *et al.* 2007, chapter 15), is required for a reliable estimate of the genetic variance in a breeding population. Once the material and the trial environment are determined, we can increase heritability by making σ_m^2 and σ_t^2 as small as possible through good experimental design and practice. This is why we make more progress through selection in well-designed field trials than in poorly designed ones.

To estimate heritability we must first estimate the additive genetic variance as well as the variance components σ_m^2 and σ_t^2. The estimation of additive genetic variance depends on the exact relationships among seedlots in the field trials. Seedlots may have been collected from a series of different mother trees in natural forests, plantations or progeny trials, or they may arise from controlled pollination. We shall consider the common situation where the seedlots are open-pollinated families. The principles are the same for more complex mating designs, but the algebra and some of the assumptions are different. For a more comprehensive treatment of genetic variance estimation see Becker (1992).

Open-pollinated families are derived from seed collected from individual seed-parents in which pollination occurs naturally and is not controlled by the breeder. If each seedling in an open-pollinated family has the same maternal parent and a different and unrelated pollen parent, the coefficient of relationship (r) of each seedling with each other seedling in the same family is 1/4 and they are strictly half-sibs, hence the term 'half-sib family'. However, we know that this rarely occurs in nature, e.g. there are likely to be seedlings in the family with the same father or with related fathers, and perhaps there will be seedlings resulting from self-pollination. In these cases the coefficient of relationship is greater than 1/4. Squillace (1974) estimated the average coefficient for open-pollinated seed from unrelated pines to be about 1/3 and, because eucalypts have a greater propensity for self-pollination, we recommend a coefficient of 1/2.5 for natural eucalypt populations (Eldridge *et al.* 1993, chapter 18). These values are important for the calculation of heritability because, for open-pollinated families, the additive variance is σ_f^2 / r where σ_f^2 is the seedlot variance component. It follows that:

$$h^2 = \frac{\sigma_f^2 / r}{\sigma_p^2} \tag{6.6}$$

Example 6.2 (continued)
We make use of the estimated variance components calculated earlier, in (6.4). Thus $\hat{\sigma}_p^2 = 3.922 + 0.276 + 0.258 = 4.456$; so, from (6.6) the estimated heritability in the population represented by the parents in the trial is:

$$\hat{h}^2 = \frac{0.258 / 0.3}{4.456} = 0.19$$

as given in Table 6.1.

Note that we have used an average coefficient of relationship among open-pollinated progeny of 0.3, because we know that: (i) *A. mangium* is insect-pollinated and there are therefore likely to be a limited number of males contributing pollen and (ii) from molecular genetics studies (Butcher *et al.* 1999) the PNG provenances of *A. mangium* have very high outcrossing rates. We can conclude that the heritability for *dbh* for the PNG provenance in this experiment at this time is 0.19, which is fairly typical for growth traits in forest trees (Eldridge *et al.* 1993, chapter 18; Cornelius 1994). How reliable is this estimate? We can calculate the standard error (s.e.) of the heritability by using a Taylor series approximation for the variance of a ratio of two variates (Kendall *et al.* 1987). Alternatively, we can use a simpler approximation that was proposed by Dickerson (1969) and supported by Dieters

Table 6.1   Genstat output for the estimation of *dbh* heritability for the PNG provenance in Example 6.2.

REML variance components analysis

Response variate:	v[1]; dbh – diameter at breast height (cm)
Fixed model:	Constant + repl + prov
Random model:	prov.family
Number of units:	164

Residual term has been added to model

Sparse algorithm with AI optimisation
Analysis is subject to the restriction on v[1]

Estimated variance components

Random term	component	s.e.
prov.family	0.258	0.129

Residual variance model

Term	Model(order)	Parameter	Estimate	s.e.
Residual	Identity	Sigma2	1.167	0.151

hmean	sigma2_t	sigma2_m	sigma2_f	h2
4.40268	3.92165	0.27596	0.25844	0.19333

et al. (1995). The Dickerson method effectively assumes that the denominator of (6.6) is a constant, leading to the approximation:

$$s.e.(h^2) \approx \frac{s.e.(\sigma_f^2)/r}{\sigma_p^2}$$

So from Table 6.1, the standard error for our estimated heritability is $\dfrac{0.129/0.3}{4.456} = 0.10$, 52% of the heritability value itself.

We shall use our estimate of h^2 to predict the response to selection in this trial. Suppose we wish to make a clonal seed orchard consisting of the best 30 individuals in the trial (*A. mangium* can be propagated by marcotting branches of the selected trees). If we consider only selection among the PNG trees, there are 748 trees left (72 trees have died), and the proportion of trees selected is $30/748 = 0.04$. Becker (1992) gives tables converting proportion selected into selection intensity in standard deviation units; a number of online sources for these tables are also readily available. A proportion selected of 0.04 corresponds to a selection intensity of about 2.15 standard deviations (note that we are assuming that the frequency distribution of *dbh* in the trial is normally distributed – this might not actually be the case). Heritability is 0.19 and the estimated phenotypic standard deviation is the square root of the phenotypic variance, i.e. $\hat{\sigma}_p = \sqrt{4.456} = 2.11$ cm. So, from (6.5) the response to selection is:

$$R = 2.15 \times 0.19 \times 2.11 = 0.86 \text{ cm}$$

This means that we would expect that at three years of age the trees derived from our 30-clone seed orchard would be 0.86 cm larger in *dbh*, on average, than those from the original population of 41 PNG families. The experimental mean of the PNG trees was 14.2 cm, so the gain in percentage terms is 6%. Note that we have not been precise in our prediction. Many factors outside our control will affect the degree of gain that will actually be achieved. The prediction depends on our estimate of heritability, which has a standard error 52% as great as its own value. It also assumes that all the selected clones will flower in synchrony and will contribute equally to the pollen pool and seed production in the seed orchard – assumptions unlikely to be realised in practice. The equation to predict genetic gain can be formulated in a simpler way (Falconer and Mackay 1996, chapter 11), namely:

$$R = h^2 S$$

where S is called the selection differential and is the difference between the mean of selected trees and the original population mean.

We have just predicted the response to selection in terms of one very simple strategy, a 30-clone orchard incorporating the best 30 individuals in the progeny test. We have made no use of the fact that the individuals are also members of families. To do this we must partition the selection effort into selection between families and selection within families. Calculations for this strategy depend on family sizes and how the selection effort is to be partitioned. We can adjust the selection effort allocated to each form of selection (within- or between-family selection) to achieve the greatest response for a given selection intensity. This is usually done through application of a combined selection index (White *et al.* 2007, chapter 15).

6.5 Heritability from provenance/progeny trials

Often, there is more structure among the seedlots than their being open-pollinated families. Seedlots can represent different provenances as well as different families, i.e. both provenance and mother-tree identity have been retained. Such trials are called provenance/progeny trials, as distinct from provenance trials. Provenance trials use provenance bulk seedlots, comprising seeds mixed together from many mother trees. It is not possible to calculate genetic parameters from provenance trials because the genetic relationships between the trees within provenances are

Eucalyptus globulus growing in a provenance trial near Kunming, China. The small trees in this plot were planted to replace trees which had died. Although popular, this practice is rarely successful and leads to problems with analysis.

not well-defined, since family identity has been lost. Although provenance trials may demonstrate significant differences between provenances, and suggest where further collections may be carried out, predictions of gain from breeding are not well-founded.

We may estimate heritability from provenance/progeny trials, but we must be careful to exclude provenance differences from the heritability. This is because one of the assumptions underlying the estimation of heritability in trials of open-pollinated families is that for each female parent, the trees used as males represent the same population. In other words, differences between families are assumed to

Provenance/progeny trial of *Acacia auriculiformis* from the Bensbach region of Papua New Guinea growing on Melville Is., Northern Territory. This photo was taken at age 3.5 years after the stand had been converted to a seedling seed orchard by thinning the original five trees to a single tree in each plot. The project is a collaborative effort between CSIRO, the NT Conservation Commission and Melville Forest Products Ltd.

be caused by differences among the female parents alone and not by systematic differences between males. If there are differences between provenances, these will be reflected not only in differences between female parents, but also among male parents. So, we must take account of the provenance differences before estimating the family variance component.

Example 6.2 (continued)

Continuing our examination of the *A. mangium* progeny trial in Example 6.2, the overall mean *dbh* of all 48 families in the trial was 13.9 cm. The mean of the two families from Sabah was 13.4 cm, and the mean of the five families from Queensland was 11.9 cm. We can remove the **restrict** command from the Genstat code in Table C6.1 and analyse the full data set of 48 families and three provenances. The estimate of heritability (using $r = 0.3$) is 0.26, somewhat larger than that obtained for the PNG provenance alone. This estimate is the pooled within-provenance heritability across all three provenances tested, with most information provided by the PNG provenance, which is best-represented in the trial. The heritabilities within the other two provenances are poorly estimated since they have only five and two families, giving inadequate samples of the average between-family differences that generate the family variance components within these provenances. Estimation is complicated by our knowledge that Queensland provenances of A. *mangium* have much higher levels of inbreeding than those from PNG (Butcher *et al.* 1999), so they should really be assigned a different coefficient of relationship, e.g. 1/2.5 (0.4), rather than 0.3. The estimated value for h^2 of 0.26 is a rough approximation only.

Note what happens if we calculate the heritability ignoring the provenance structure in the seedlots by omitting the term *prov* from the list of fixed terms in the **vcomponents** line of Table C6.1. The estimated heritability becomes 0.58, greatly inflated as a result of incorrectly incorporating into the calculation the substantial differences between provenances evident in this trial.

When predicting the genetic gain from selection in provenance/progeny trials, we must take into account both the gain from selecting between provenances, and the gain from selecting between families and individuals within provenances. In Example 6.2, the genetic gain from the clonal seed orchard would be higher relative to the overall mean *dbh* (13.9 cm) than relative to that of the best provenance, PNG (14.2 cm). However, in this case, given the prior expectation that PNG would outperform the other provenances, gain relative to the PNG provenance is the most relevant consideration. Genetic gain from selection between provenances can be predicted using provenance heritability (Burley and Wood 1976; Otegbye and Samarawira 1992), which is a totally different concept from narrow-sense additive heritability as defined above, and is better termed 'provenance repeatability'. We will simply note here that selection of the best provenances is fundamentally important in breeding.

Uncertainties about the coefficient of relationship within families and how to account for provenance effects in open-pollinated provenance/progeny trials mean that genetic parameters and predicted genetic gains cannot be precisely estimated from such trials. Inbreeding reduces vigour in many forest tree species, so varying levels of inbreeding in different families further complicates the calculation of genetic parameters for growth traits in open-pollinated trials (Griffin and Cotterill 1988). The rankings of parents for their breeding value and estimates of heritability are much more reliable after one generation of control-pollinated breeding, because variation in the breeding system is eliminated and provenance effects can be better accounted for (Hodge *et al.* 1996).

Molecular genetics has opened the possibility of directly determining the genetic relatedness among individuals within and between families, using genomic markers. This information can be incorporated into the estimation of heritability and other genetic parameters. For example, Bush *et al.* (2011) used breeding system estimates derived from isozyme analysis to adjust outcrossing rates in different provenances of *Eucalyptus cladocalyx* and reported that heritability estimates for this species were markedly reduced when these varying outcrossing rates were accounted for, as some natural provenances displayed high levels of inbreeding.

There are levels of hierarchy higher than provenance which we should take into account wherever possible. As an example, the natural distribution of *Eucalyptus camaldulensis* covers most of Australia. Based on molecular and morphological evidence, McDonald *et al.* (2009) described seven subspecific taxa within this species. The main subspecies native to southern Australia, subsp. *camaldulensis* is planted as an exotic in temperate regions while two important northern taxa, subsp. *obtusa* and subsp. *simulata* are widely planted in the wet–dry tropics. If provenances from these three subspecies were planted in the same provenance trial, provenance differences would include the major differences between subspecies, which if not accounted for would overestimate the provenance differences to be found within each. Similarly, there are major discontinuities in the natural distributions of tropical acacias, such as *Acacia auriculiformis*, *A. crassicarpa* and *A. mangium*, caused by the separation of Queensland from PNG by Torres Strait; these are reflected in regional differences in provenance performance. It is, however, unsafe to assume that a physical discontinuity such as Torres Strait will necessarily lead to discontinuities in provenance variation. It is also unsafe to assume that close proximity of provenances will always mean similarity (Matheson and Mullin 1986). The point here is to explicitly identify known sources of variation wherever possible and take them into account by declaring them to be fixed effects before proceeding with the analysis.

6.6 Genetic correlation

Genetic correlation between two traits is an expression of how much change we expect in one trait when selecting for the other. The genetic correlation (r_{xy}) between two traits x and y is defined as:

$$r_{xy} = \frac{\sigma^2_{xy}}{\sqrt{\sigma^2_x \sigma^2_y}} \tag{6.7}$$

where σ^2_{xy} is the covariance component for x and y, and σ^2_x and σ^2_y are variance components. Genetic correlation is used mainly for four different purposes:

1. To help predict response at harvest to selection carried out in young trees. We very rarely estimate heritabilities for traits at harvest because of the delay this would cause to the breeding program. We carry out evaluation and selection much earlier than this and usually assume that the genetic correlation between what we measure and what we want is high.

2. To predict response in a trait which is hard to measure from one which is easy to measure. Wood density takes much time to measure accurately and involves

some damage to the trees being measured. Shooting a nail into a tree with a pilodyn and measuring its penetration is simple, quick and involves little damage to the tree. Although there is a positive genetic correlation with wood density, it is not perfect (Raymond and MacDonald 1998).

3. To predict response to selection at one site when selecting at another. Growth at one site may be regarded as one trait and growth at another site may be regarded as another trait. The higher the genetic correlation, the less the genotype-by-environment interaction (Burdon 1977).

4. Selection indices are constructed using genetic correlations and heritabilities to maximise gain in specified traits selected at the same time.

We calculate genetic correlation from progeny trials in much the same way as heritability, estimating family variance (and covariance) components and calculating their appropriate function. As for the standard product-moment correlation, genetic correlation between two traits is the ratio of genetic covariance to the pooled estimate of the genetic variance. If we have measurements on traits x and y then we make use of the relation $(x + y)^2 = x^2 + y^2 + 2xy$ to obtain an estimator for the covariance component σ^2_{xy} between two traits. We can analyse x and y separately and the sum $(x + y)$ and use the above expression to obtain:

$$\hat{\sigma}^2_{xy} = \frac{1}{2}(\hat{\sigma}^2_{(x+y)} - \hat{\sigma}^2_x - \hat{\sigma}^2_y)$$

where $\sigma^2_{(x+y)}$ is the variance component for the sum of the traits.

This method provides a convenient mechanism using our already established techniques. However, its disadvantages are that it does not provide a standard error and it is cumbersome if we wish to analyse the correlations among many traits. Then it is better to specify a multivariate mixed-model procedure to estimate genetic correlations and their standard errors (Apiolaza *et al.* 2000). The model specification and analyses are not straightforward and are not covered in this book.

Example 6.2 (continued)

If we now consider both the measured variates in this experiment, namely *dbh* and *ht*, we can calculate plot means for each variate as well as for the derived variate *sum*, obtained by adding the *dbh* and *ht* measurements for each tree. The data file of plot means is given in Table B6.2. We can then analyse these variates (*dbh*, *ht* and *sum*) separately in order to estimate genetic correlation between *dbh* and *ht* at three years of age and to estimate gain in *ht* resulting from selection for *dbh*. As with heritability, we will examine only the 41 PNG families by restricting the analysis to that provenance.

In Section 6.2, we needed to estimate σ^2_m, σ^2_t and σ^2_f for the calculation of heritability, thus the full plot summary file of plot means, variances and counts was required in Table B6.1. To calculate the genetic correlation between *dbh* and *ht*, however, we just need estimates of σ_f^2 for *dbh* (*x*), *ht* (*y*) and *sum* (*x* + *y*). The Genstat program to estimate the genetic correlation is given in Table C6.2 and the relevant output in Table 6.2. The estimated seedlot variance components for *dbh*, *ht* and *sum* are 0.258, 0.271 and 0.873 respectively. The required covariance component is then $(0.873 - 0.258 - 0.271)/2 = 0.172$. Hence, from (6.7) the genetic correlation is estimated as $\hat{r}_{xy} = 0.65$.

The response in trait *y* to selection in trait *x* is called the correlated response (*CR*) to selection and is defined as:

$$CR = i h_x h_y r_{xy} \sigma_{py} \tag{6.8}$$

where h_x and h_y are the square roots of the two heritabilities and σ_{py} is the phenotypic standard deviation in trait *y*. These correlated responses are particularly important for index selection. They are also important in cases where we cannot measure exactly the same trait as the one we wish to improve. For example, we cannot measure merchantable volume without felling and processing the tree; we usually measure standing volume instead. Nor do we usually measure volume at harvest age; we measure volume at an earlier age. Although it is rarely done, breeders should carry out an experiment to estimate the genetic correlations between desired traits and measurable traits, and use these genetic correlations in gain prediction equations.

Table 6.2 Part of the Genstat output for the estimation of genetic correlation between *dbh* and *ht* for the PNG provenance in Example 6.2.

REML variance components analysis

Response variate:	v[1]; dbh – diameter at breast height (cm)
Fixed model:	Constant + repl + prov
Random model:	prov.family
Number of units:	164

Estimated variance components

Random term	component	s.e.
prov.family	0.258	0.129

Residual variance model

Term	Model(order)	Parameter	Estimate	s.e.
Residual	Identity	Sigma2	1.167	0.151

Response variate:	v[2]; ht – height (m)
Fixed model:	Constant + repl + prov
Random model:	prov.family
Number of units:	164

Residual term has been added to model

Sparse algorithm with AI optimisation
Analysis is subject to the restriction on v[2]

Estimated variance components

Random term	component	s.e.
prov.family	0.271	0.124

Residual variance model

Term	Model(order)	Parameter	Estimate	s.e.
Residual	Identity	Sigma2	1.055	0.136

Response variate:	v[3]; sum – sum = dbh + ht
Fixed model:	Constant + repl + prov
Random model:	prov.family
Number of units:	164

Estimated variance components

Random term	component	s.e.
prov.family	0.873	0.391

Residual variance model

Term	Model(order)	Parameter	Estimate	s.e.
Residual	Identity	Sigma2	3.244	0.419

sigma2_f[1]	sigma2_f[2]	sigma2_f[3]	gencor
0.25844	0.27111	0.87275	0.64828

For our example to predict gain in *ht* from a clonal seed orchard of the 30 individuals of greatest *dbh*, we need the intensity of selection (2.15 as before), heritability for *dbh* (0.19), heritability for *ht* (0.31), and the phenotypic standard deviation of *ht*, 2.93 m (the calculations of heritability and phenotypic standard deviation for *ht* are not shown here). Using (6.8), the correlated response in *ht* from selection on *dbh* is calculated as $2.15 \times \sqrt{0.31} \times \sqrt{0.19} \times 0.65 \times 2.93 = 0.99$ m.

This can be compared with the response to direct selection for *ht* of $2.15 \times 0.31 \times 2.93 = 1.95$ m. In this example, selecting for *dbh* delivers about half the genetic gain in *ht* that would be obtained by selecting directly for *ht*. The fact that the heritability for *dbh* is substantially lower than that for *ht* reduces the correlated response to some extent, relative to the results of direct selection for *ht*.

These calculations tell us that we could expect considerable response in *ht* to selection for *dbh* in the above example. This is expected wherever the genetic correlation between the traits is strongly positive. Some estimates of genetic correlation are negative; selection for increase in one trait leads to reduction in another. In such cases the seedlot covariance component (and the stratum covariance) is negative, alarming but not unusual. An example of this is growth rate and wood density in pines, where most estimates of genetic correlation are negative. However, the standard errors of genetic correlations are usually quite large and it is unwise to rely too heavily on the actual numerical values calculated. It is more common to rely on the sign of genetic correlations than their exact value, particularly when several studies give similar results from different material.

Genetic correlations have some influence on the best way to structure breeding populations. Where there are strongly negative genetic correlations between two traits, it is very difficult to breed for both in the same population. That would be like selecting in two directions at once. This is called 'disruptive selection' and it leads to higher genetic variance but little change in the mean of the breeding population. In such cases it is better to divide the material into two populations and select separately with emphasis on only one trait in each population. Crossing between the populations later allows us to select the very few individuals ('correlation breakers') that combine both traits in a desirable way. Such individuals might be propagated clonally and information from their parentage can be used to select for further breeding in each parent population. An alternative approach is to select for one trait, but using a restriction index (Kempthorne and Nordskog 1959) to ensure that there is no or little negative change in the other. Where there are strong positive genetic correlations, it is not necessary to select for all traits. Some will piggyback on selection for others and a subdivided population will not be necessary.

Where there is strongly positive genetic correlation between a trait in young trees and the same trait when older, it is probably not necessary to go to the expense of the later measurements. Selection can be made earlier and breeding can proceed more quickly. Such genetic correlations are often called 'juvenile–mature correlations' even though they have little to do with the maturation state of the trees in question. A better term is 'age–age correlations' (Lambeth 1980).

6.7   Selecting superior treatments – BLUEs versus BLUPs

The mixed-effects model (6.3) estimates the variance component for treatments; in so doing it generates BLUPs for each treatment, and these are very useful. Modern practice in tree breeding (White *et al.* 2007, chapter 15) is to rank seedlots, clones and individual trees using their BLUPs when selecting trees for further breeding or deployment.

Here, we simply demonstrate the output of family BLUPs for the 41 PNG families in Example 6.2, and compare them with the corresponding BLUEs obtained if we specify *family* as a fixed factor in the linear model (Table 6.3). Examining this table one can see that the range of BLUPs for the families is slightly shrunk towards the grand mean (14.2 cm), relative to that of the BLUEs. In general, in trials where the genetic treatment variance component is small relative to the residual variance this shrinkage is greater. As the treatment variance component approaches zero, the range of the treatment BLUPs will also decline towards zero; whereas the range of the BLUEs will not. Treatments which are poorly represented have their BLUPs regressed further towards the grand mean. Selecting on BLUPs rather than BLUEs therefore has the advantage that genetic treatments (families, individual trees, clones etc.) will not be chosen for further breeding or deployment based on their good performance if they are represented in only one or a very few plots. BLUPs can be determined for a multi-trait breeding objective, taking into

Table 6.3   BLUE and BLUP *dbh* values for PNG families in Example 6.2.

Family number	BLUE	BLUP
1	14.99	14.58
2	15.05	14.60
3	13.25	13.76
4	13.97	14.10
5	15.89	15.00
6	13.31	13.79
8	15.07	14.62
9	14.12	14.17
10	15.21	14.68
11	13.53	13.89
12	13.39	13.83
13	14.35	14.28
15	12.49	13.40
16	14.66	14.42
17	13.01	13.65
18	13.63	13.94
19	13.61	13.93
20	13.66	13.95
22	14.55	14.37
23	15.53	14.83
25	14.85	14.51

Table 6.3 Continued

26	14.27	14.24
27	14.37	14.29
29	14.56	14.38
30	14.35	14.27
31	13.49	13.87
32	14.55	14.37
33	14.78	14.48
34	14.62	14.40
36	13.55	13.90
37	13.82	14.03
38	13.82	14.03
39	14.33	14.27
40	13.41	13.83
41	14.48	14.34
42	13.93	14.08
43	14.38	14.29
44	14.83	14.50
45	14.07	14.15
46	13.39	13.83
47	15.39	14.76

account the economic weightings given to different traits, and genetic parameters such as heritabilites and genetic correlations among the traits.

Finally, we note that progeny trials can be analysed using an individual tree model (White *et al.* 2007, chapter 15). Ideally, an initial analysis of the plot summary file (Chapter 3) should first be undertaken, in order to effectively screen the data. Genetic relationships among the individual trees must be specified using a pedigree file. Specification of genetic relationships may be informed by molecular genetic information (Bush *et al.* 2015). Such an analysis can generate BLUP breeding values for all the individual trees in the trial, and for the parent trees from which the seedlots were collected. A joint analysis of two or more trials that have genetic linkage (i.e. common genetic treatments) can generate BLUPs for all trees in the trials. This type of analysis is not straightforward and is beyond the scope of this book. Individual-tree models are now widely used in advanced tree breeding, since they enable the generation of breeding values across multiple trials and multiple generations in 'rolling front' breeding programs, facilitating more rapid genetic improvement; see for example FWPA (2022).

7

Incomplete block designs

7.1 Introduction

In Chapter 4 we discussed some simple experimental designs such as the RCB design, the two-factor design and the split-plot design. Analysis of these designs is straightforward because the treatment factors are orthogonal to the blocking factors. To explain what is meant by orthogonality of factors, consider an RCB design for r replicates and v seedlots. As the name suggests, the design is a complete block design, i.e. in each replicate (the blocking factor) there is a single representative of each of the v seedlots. We say that the blocking factor *repl* is orthogonal to the treatment factor *seedlot*.

In this chapter we will look at non-orthogonal designs, an example of which is the incomplete block design, i.e. one where not all representatives of v treatments appear in each block. We will explain why it is advisable to use incomplete block designs for many tree improvement field trials. We will introduce a class of non-orthogonal designs known as generalised lattice designs and describe a computer software package called CycDesigN which simplifies the construction and randomisation of this more complex class of experimental design. We will also discuss developments in the construction of (i) spatial designs; these can be viewed as enhancements to the more traditional use of incomplete blocks and (ii) partially replicated designs for use in initial screening trials.

7.2 Need for incomplete block designs

First, let us look at an RCB design for three replicates and 42 treatments. A possible field layout (after randomisation) is given in Fig. 7.1. As discussed in Section 4.2.2, the model for an RCB design includes replicate parameters to allow adjustment to accommodate differences between the replicates. A consequence of the RCB model, however, is that we assume all comparisons between plots within a replicate are made at the same level of precision regardless of the physical distance between plots. If the number of treatments is small, this assumption is reasonable, when supported by randomisation of treatments to plots. As the number of treatments increases, however, the number of plots in each replicate gets larger and consequently the maximum distance between plots within replicates increases. Therefore, the assumption that the between-plot variance is the same for any two plots in a replicate becomes less reliable, and so it is less satisfactory to make direct comparisons between treatment means within a replicate.

For example, the plots in Fig. 7.1 might represent 25-tree plots at 2×2 m spacing, so each plot would be 0.01 ha and a replicate would be 0.42 ha. This is a large area within which to make the assumption (inherent in the RCB model) of equal pairwise variances between plots within replicates. It would be much better to subdivide replicates so that, if necessary, an adjustment can be made within replicates (as well as between replicates) to more adequately account for field variation and thus allow a more accurate comparison of treatment means. Subdividing replicates into smaller components introduces the concept of 'incomplete blocks'. The incomplete blocks (or just blocks) are the components into which we have divided each replicate. We could think of the columns of each replicate in Fig. 7.1 as constituting incomplete blocks. The first block in replicate 1

42	12	15	37	17	22	4	
21	26	25	18	28	36	16	
10	35	6	29	11	8	2	Replicate 1
23	38	41	3	31	24	30	
39	1	27	40	14	34	7	
19	20	33	5	9	13	32	
25	16	5	6	29	40	24	
23	33	21	42	22	19	41	
12	36	8	34	32	15	39	Replicate 2
2	20	28	35	26	13	3	
9	14	1	7	10	4	30	
18	37	27	17	31	38	11	
33	16	21	7	1	41	37	
17	28	40	29	24	36	35	
4	26	34	11	10	18	39	Replicate 3
22	19	9	13	14	31	25	
5	6	20	12	2	38	32	
23	3	30	27	15	42	8	

Fig. 7.1. An RCB design for three replicates of 42 treatments.

would contain the treatments 42, 21, 10, 23, 39 and 19. The blocks are called incomplete because not all the treatments appear in each block; in our example only six out of the possible 42 treatments appear in each block. The extra blocking structure introduced to further control field variation appears as an extra term in the model, which now becomes:

$$(\text{observation}) = (\text{overall mean}) + (\text{replicate effect}) + $$
$$(\text{block effect}) + (\text{treatment effect}) + (\text{residual})$$

or symbolically:

$$Y_{igh} = \mu + \rho_i + \beta_{ig} + \tau_j + \varepsilon_{igh} \qquad (7.1)$$

where the Y_{igh} ($i = 1,2,...,r$; $g = 1,2,...,s$; $h = 1,2,...,k$) are the observations for r replicates each with s incomplete blocks containing k plots; μ is a parameter for the overall mean; the ρ_i, β_{ig} and τ_j are parameters for replicates, incomplete blocks and treatments respectively; and the ε_{igh} are residuals in the model. To explain further why it is a good idea to have incomplete blocks in field trials where the number of treatments is large, e.g. more than 16, let us suppose that in replicate 1 of Fig. 7.1 there is a field trend such that block 1 is on low ground and block 7 is on high ground, as shown in Fig. 7.2.

Often, plots on lower ground have greater soil depth and soil water availability than those on higher ground, and thus support faster tree growth. Thus it would be unreasonable to compare the mean of treatment 19 in block 1

				Block			
1	2	3	4	5	6	7
42	12	15	37	17	22	4
21	26	25	18	28	36	16
10	35	6	29	11	8	2
23	38	41	3	31	24	30
39	1	27	40	14	34	7
19	20	33	5	9	13	32

Low ground ——————————————→ High ground

direction of trend

Fig. 7.2. Replicate 1 of Fig. 7.1 subdivided to illustrate the use of incomplete blocks to adjust for a trend in field conditions.

with that of treatment 32 in block 7. We would, however, feel happier about comparing the mean of treatment 19 with that of treatment 42. So, by organising the incomplete blocks to run at right angles to the direction of the trend we can divide the comparisons between means into those that are reasonable (within blocks) and those that are less desirable (between blocks). With the incomplete block model we can provide a parameter for each block that will make an adjustment from block to block depending on the nature of the field variation. Hence, the estimated treatment means from the model can all be compared.

The introduction of block parameters, however, means that between-treatment comparisons are made more accurately for pairs of treatments in the same block. The aim of the statistician is to design the experiment so that all pairs of treatments can be compared with roughly the same precision. This would not happen if, e.g., treatments 1 and 10 were in the same block in each replicate of a design. There would be high precision (low standard error) for the comparison of treatments 1 and 10, but then other pairwise treatment

Plantings where there is a well-defined environmental gradient require special attention. This *Acacia* species trial in northern Guangdong Province, China, has 25-tree square plots and was intended to compare 44 seedlots in four replicates. The slope causes the experiment to be quite different at the top of a replicate from the bottom. If incomplete blocks were included in the design (at right angles to the slope), these effects could be taken into account. (Photo Suzette Searle)

comparisons would be disadvantaged. Notice that in Fig. 7.1 no two treatments appear together within blocks more than once. That is, any chosen pair of treatments will either appear in a block just once, or not at all. This is no accident – the layout in Fig. 7.1 has been constructed to have this property. It would be unlikely to happen if the 42 treatment numbers were randomly allocated to the plots of each replicate, as we would do with an RCB design. We now describe how to produce a good design, in which pairs of treatments appear together within blocks roughly equally.

7.3 Choice of incomplete block design

Consider an RCB design for v treatments and r replicates. We want to break the v plots in each replicate into s incomplete blocks of size k plots, so therefore we have $v = sk$.

There are many ways that treatments can be arranged into blocks in each replicate; some make more sense than others. For example, in Fig. 7.3 there are two different arrangements for a design with $r = 2$, $v = 9$ and $k = 3$.

(a)

	Replicate 1			Replicate 2		
Block	1	2	3	1	2	3
	1	4	7	1	2	3
	2	5	8	4	5	6
	3	6	9	7	8	9

(b)

	Replicate 1			Replicate 2		
Block	1	2	3	1	2	3
	1	4	7	1	5	4
	2	5	8	2	8	6
	3	6	9	3	9	7

Fig. 7.3. Two possible arrangements for an incomplete block design with r = 2, v = 9 and k = 3.

Which of these arrangements is the most desirable? Later we will develop a way to measure the various possibilities so that the optimal or near-optimal design can be chosen. For the moment, we can argue heuristically. We assume treatments can be compared more accurately if they appear together somewhere in a block. If all the treatments have equal status, i.e. before the experiment we are not more likely to want to compare treatments A and B than to compare treatments A and C, it is sensible to organise the treatments in blocks so that the maximum number of different pairwise comparisons between treatments occur within blocks. In Fig. 7.3(a) treatment 1 appears with treatments 2 and 3 in replicate 1 and treatments 4 and 7 in replicate 2. So the pairwise comparisons available with treatment 1 are (1,2) and (1,3), (1,4) and (1,7). But in Fig. 7.3(b) the only comparisons are (1,2) and (1,3). So the design in Fig. 7.3(a) is more desirable than that in Fig. 7.3(b) as far as treatment 1 is concerned, and in fact this is the case for all the treatments. Thus we would prefer to have the arrangement of treatments in Fig. 7.3(a) than in Fig. 7.3(b), but maybe there is a better arrangement still. Clearly, what is required is some way of comparing designs so that, for a fixed number of replicates, treatments and block sizes, we can choose the most desirable allocation of treatments. A quantity called the average efficiency factor allows us to make this comparison. First, we look at parameter estimation and develop some notation.

Parameter estimation is more complicated with the incomplete block model (7.1) than with the models described in earlier chapters because blocks and treatments are not orthogonal, i.e. not all the treatments appear in each block. So the estimation process has to calculate adjustments to raw treatment means to take account of which blocks the treatments are in. This is why we have included blocks in the model; the computer can easily handle the more complicated calculations, but to explain what is going on we will specify the model in more formal terms. It is convenient to write the model in terms of matrices.

If Y is a vector comprising all the individual observations Y_{igh}, and likewise ρ, β and τ are vectors of parameters for replicate, block and treatment effects, then the model (7.1) can be written as:

$$Y = G\mu + R\rho + Z\beta + X\tau + \varepsilon$$

where R, Z and X are called the design matrices for replicates, blocks and treatments respectively and G is a vector of n ones ($n = rsk$). To illustrate, look at Fig. 7.3(a) and suppose the order of the observations in Y is replicate 1, block 1, plot 1; replicate 1, block 1, plot 2 etc., down to replicate 2, block 3, plot 3. Then:

$$R = \begin{bmatrix} 1 & 0 \\ 1 & 0 \\ 1 & 0 \\ 1 & 0 \\ 1 & 0 \\ 1 & 0 \\ 1 & 0 \\ 1 & 0 \\ 1 & 0 \\ 0 & 1 \\ 0 & 1 \\ 0 & 1 \\ 0 & 1 \\ 0 & 1 \\ 0 & 1 \\ 0 & 1 \\ 0 & 1 \\ 0 & 1 \end{bmatrix} \quad Z = \begin{bmatrix} 1 & 0 & 0 & 0 & 0 & 0 \\ 1 & 0 & 0 & 0 & 0 & 0 \\ 1 & 0 & 0 & 0 & 0 & 0 \\ 0 & 1 & 0 & 0 & 0 & 0 \\ 0 & 1 & 0 & 0 & 0 & 0 \\ 0 & 1 & 0 & 0 & 0 & 0 \\ 0 & 0 & 1 & 0 & 0 & 0 \\ 0 & 0 & 1 & 0 & 0 & 0 \\ 0 & 0 & 1 & 0 & 0 & 0 \\ 0 & 0 & 0 & 1 & 0 & 0 \\ 0 & 0 & 0 & 1 & 0 & 0 \\ 0 & 0 & 0 & 1 & 0 & 0 \\ 0 & 0 & 0 & 0 & 1 & 0 \\ 0 & 0 & 0 & 0 & 1 & 0 \\ 0 & 0 & 0 & 0 & 1 & 0 \\ 0 & 0 & 0 & 0 & 0 & 1 \\ 0 & 0 & 0 & 0 & 0 & 1 \\ 0 & 0 & 0 & 0 & 0 & 1 \end{bmatrix} \quad X = \begin{bmatrix} 1 & 0 & 0 & 0 & 0 & 0 & 0 & 0 & 0 \\ 0 & 1 & 0 & 0 & 0 & 0 & 0 & 0 & 0 \\ 0 & 0 & 1 & 0 & 0 & 0 & 0 & 0 & 0 \\ 0 & 0 & 0 & 1 & 0 & 0 & 0 & 0 & 0 \\ 0 & 0 & 0 & 0 & 1 & 0 & 0 & 0 & 0 \\ 0 & 0 & 0 & 0 & 0 & 1 & 0 & 0 & 0 \\ 0 & 0 & 0 & 0 & 0 & 0 & 1 & 0 & 0 \\ 0 & 0 & 0 & 0 & 0 & 0 & 0 & 1 & 0 \\ 0 & 0 & 0 & 0 & 0 & 0 & 0 & 0 & 1 \\ 1 & 0 & 0 & 0 & 0 & 0 & 0 & 0 & 0 \\ 0 & 0 & 0 & 1 & 0 & 0 & 0 & 0 & 0 \\ 0 & 0 & 0 & 0 & 0 & 0 & 1 & 0 & 0 \\ 0 & 1 & 0 & 0 & 0 & 0 & 0 & 0 & 0 \\ 0 & 0 & 0 & 0 & 1 & 0 & 0 & 0 & 0 \\ 0 & 0 & 0 & 0 & 0 & 0 & 0 & 1 & 0 \\ 0 & 0 & 1 & 0 & 0 & 0 & 0 & 0 & 0 \\ 0 & 0 & 0 & 0 & 0 & 1 & 0 & 0 & 0 \\ 0 & 0 & 0 & 0 & 0 & 0 & 0 & 0 & 1 \end{bmatrix}$$

Using least squares theory, the treatment effects are estimated by solving the following equations, which are known as the reduced normal equations:

$$(rI - \frac{1}{k}NN')\hat{\tau} = X'Y - \frac{1}{k}NZ'Y \tag{7.2}$$

where $\hat{\tau}$ is the vector of estimated treatment effects, X' is the transpose of X, I is the $v \times v$ identity matrix and $N = X'Z$ (John and Williams 1995, chapter 1).

Here we want to concentrate on two important matrices. First, N is called the incidence matrix of the design: it is a $v \times rs$ matrix whose (j,ig)th element is the number of times treatment j appears in block g of replicate i of the design. Normally it is desirable for treatments to appear at most once in any block; N will then consist of zeros and ones. The matrix NN' is called the concurrence matrix of the design and it is the properties of this matrix which determine the desirability of a particular design. The (j,j')th element of the concurrence matrix is the number of times treatments j and j' appear together within blocks of the design. Hence the diagonal elements of the concurrence matrix are just the replication numbers for each of the treatments. As an example, the concurrence matrix for the design in Fig. 7.3(a) is

$$NN' = \begin{bmatrix} 2 & 1 & 1 & 1 & 0 & 0 & 1 & 0 & 0 \\ 1 & 2 & 1 & 0 & 1 & 0 & 0 & 1 & 0 \\ 1 & 1 & 2 & 0 & 0 & 1 & 0 & 0 & 1 \\ 1 & 0 & 0 & 2 & 1 & 1 & 1 & 0 & 0 \\ 0 & 1 & 0 & 1 & 2 & 1 & 0 & 1 & 0 \\ 0 & 0 & 1 & 1 & 1 & 2 & 0 & 0 & 1 \\ 1 & 0 & 0 & 1 & 0 & 0 & 2 & 1 & 1 \\ 0 & 1 & 0 & 0 & 1 & 0 & 1 & 2 & 1 \\ 0 & 0 & 1 & 0 & 0 & 1 & 1 & 1 & 2 \end{bmatrix}$$

From NN' we can derive a quantity known as the average efficiency factor (E) of a design (John and Williams 1995, chapter 2). The average efficiency factor is fundamental to assessing one design relative to other designs with the same basic parameters, r, v and k, and is related to the average of all the pairwise variances for treatments. When the concurrence of pairs of treatments is high the pairwise variance is low, but if pairs of treatments do not appear together within blocks at all the pairwise variance will be higher. The average pairwise variance will be minimised and the average efficiency factor maximised if all the concurrences of pairs of treatments are as similar as possible, namely if the off-diagonal elements of NN' are as similar as possible. In the above example the concurrences are either zero or one and the design is known as a (0,1)-design. The example in Fig. 7.3(b) is a (0,1,2)-design and so is less desirable because the average efficiency factor will be lower.

In general, for fixed r, v and k we want to obtain the design that has an average efficiency factor as high as possible. We can compare the average efficiency factor with a known theoretical upper bound; if the difference is slight, we say that the design is efficient. We now introduce a class of experimental designs providing a wide range of efficient designs suitable for use in practice.

7.4 Generalised lattice designs and alpha designs

7.4.1 Description

It is common in tree improvement field trials to have a large number of treatments and a small number of replicates. Suppose we have r replicates of $v = sk$ treatments. Within each replicate there are s blocks each of size k plots. There is thus a two-stage structure for the control of field variation. First, replicates can allow an adjustment for large-scale variation, then within each replicate the blocks provide a second level of adjustment.

An incomplete block design in which blocks can be grouped into replicates of the treatments is called a resolvable design. Non-resolvable designs exist but are less valuable for field trials because they do not allow a two-stage removal of field trend where, importantly, the first-stage removal is carried out by the replicates which are orthogonal to treatments. The overall structure of an r-replicate resolvable design for v treatments with s blocks of size k within each replicate is an example of a generalised lattice design. Most textbooks on experimental design (e.g. Cochran and Cox 1957) give special cases of such designs. The most common are the square lattice designs ($k = s$) and the rectangular lattice designs ($k = s - 1$). For these designs the concurrence matrices contain either zeros or ones in the off-diagonal positions and so, following the discussion in the previous section, we know that the designs will be efficient; in fact, square and rectangular lattice designs are optimal. Efficient designs are also available for other values of r, s and k. A class of generalised lattice designs, called alpha designs, caters for most situations encountered in practice (Patterson and Williams 1976). A detailed discussion of alpha designs is given in John and Williams (1995, chapter 4).

7.4.2   Construction of alpha designs

An alpha design is generated from an alpha array, which is an $r \times k$ array of numbers in the range $0,1,...,s - 1$. For example, for $r = 3$, $s = 5$ and $k = 4$ we can have the alpha array:

$$
\begin{array}{cccc}
0 & 0 & 0 & 0 \\
0 & 1 & 2 & 3 \\
0 & 2 & 4 & 1
\end{array}
\tag{7.3}
$$

The alpha design is obtained as follows:

1. Write the treatment numbers 1 to 20 in $k = 4$ groups of $s = 5$, that is:

$$
\begin{aligned}
\text{group } 1 &= 1,\ 2,\ 3,\ 4,\ 5 \\
\text{group } 2 &= 6,\ 7,\ 8,\ 9,\ 10 \\
\text{group } 3 &= 11,\ 12,\ 13,\ 14,\ 15 \\
\text{group } 4 &= 16,\ 17,\ 18,\ 19,\ 20
\end{aligned}
\tag{7.4}
$$

2. The ith row of the alpha array is used for the ith replicate of the alpha design and tells us how to rotate elements in the above groups in order to construct the design.
3. Because all elements in the first row of the alpha array are zero, the first replicate in Fig. 7.4 contains treatment numbers in natural order, and the columns represent incomplete blocks. It is usual (but not essential) for the first row of the alpha array to be zeros.

4. The second row of the alpha array tells us to:
   i   leave the first group alone (0)
   ii   cycle the second group by one place (1)
   iii cycle the third group by two places (2)
   iv   cycle the fourth group by three places (3)
   and hence the second replicate in Fig. 7.4 is obtained.
5. The third row of the alpha array tells us to:
   i   leave the first group alone (0)
   ii   cycle the second group by two places (2)
   iii cycle the third group by four places (4)
   iv   cycle the fourth group by one place (1)
   and hence the third replicate in Fig. 7.4 is obtained.

Block

1	2	3	4	5
1	2	3	4	5
6	7	8	9	10
11	12	13	14	15
16	17	18	19	20

Replicate 1

Block

1	2	3	4	5
1	2	3	4	5
7	8	9	10	6
13	14	15	11	12
19	20	16	17	18

Replicate 2

Block

1	2	3	4	5
1	2	3	4	5
8	9	10	6	7
15	11	12	13	14
17	18	19	20	16

Replicate 3

Fig. 7.4. Unrandomised alpha design for r = 3, s = 5 and k = 4.

From an alpha array, properties of the design can be determined, such as the efficiency factor. For a particular set of r, s and k many alpha arrays are possible, so we must choose one that leads to an efficient alpha design. In Section 7.6 we discuss the design generation package CycDesigN which does this for us.

Inspection of blocks in Fig. 7.4 reveals that some pairs of treatments appear together within blocks once, e.g. treatments 2 and 7, whereas other pairs do not appear at all within blocks, e.g. treatments 2 and 16. In fact all pairs of treatments appear together within blocks either zero times or once and the concurrence matrix NN' will only have zeros and ones off-diagonal. An alpha design with concurrences of only zero or one is called an alpha(0,1) design.

When the number of plots per block (k) is greater than the number of blocks per replicate (s), it is not possible to construct alpha(0,1) designs. The next best thing is to restrict the number of concurrences to two, leading to alpha(0,1,2) designs.

7.4.3 Randomisation for alpha designs

For an alpha design, there are three strata, namely *repl*, *repl.block* and *repl.block.plot*. In Genstat syntax this would be:

$$\text{block} \qquad \text{repl / block / plot} \qquad (7.5)$$

The randomisation consistent with these strata is as follows:

1. Randomise the order of replicates.
2. Randomise the order of blocks within each replicate.
3. Randomise the order of plots within each block.
4. Allocate treatments at random to the numbers 1 to v.

Where there are control treatments or there is a structure on the treatments such as groupings into say, provenances and families, restrictions may be placed on the randomisation in (4) to improve the properties of the design. One of the features of alpha designs is that nested structures, such as families within provenances, can be incorporated easily into the design. For example, suppose that the four groups defined in (7.4) actually represent four provenances, each with five families per provenance. By carrying out a restricted treatment randomisation (i.e. randomising the treatment numbers in each group separately then randomising the allocation of groups to the four provenances) we can guarantee that, in Fig. 7.4, each block will contain each provenance exactly once. Under the reasonable assumption that provenance differences will be greater than family-within-provenance differences, it is most desirable to have the maximum number of provenances represented in each block. In Fig. 7.4 provenances are in fact orthogonal to blocks.

The other extreme (which could happen with a randomisation that ignores the treatment structure) would be for one block to have treatments from only one provenance say, and a second block to have treatments from another provenance. This would not be a good idea. We introduced blocks to control field variation, so we would want to avoid mixing up or confounding provenance differences with block differences.

CycDesigN (Section 7.8) has facilities for restricted randomisation. It is worth emphasising that the above randomisation for blocking factors assumes that replicates are separate entities, and that there is no association between blocks in different replicates and, likewise, between plots in different blocks. So, although we write the design in a rectangular arrangement of plots in Fig. 7.4, the actual layout of the alpha design could have replicates, and blocks within each replicate, oriented in any way dictated by site conditions. The only real requirement is that blocks within each replicate form a physical entity for the removal of field trend. Equally, plots in each block would be grouped together to allow adjustment for field trend, but there is no requirement for them to be in a column as depicted in Fig. 7.4. In Section 7.5 we describe generalised lattice designs for more structured field layouts.

7.4.4   Availability of alpha designs

An advantage of alpha designs is that they are available for a wide range of combinations of parameters. There is no need to restrict resolvable incomplete block designs to the square and rectangular lattice designs discussed by Cochran and Cox (1957, chapter 10). Efficient designs can be constructed for all combinations of r, s and k that would be required in practice. CycDesigN makes these designs available.

For some numbers of treatments there may be a choice of block sizes, e.g. an alpha design for $v = 48$ treatments could have block sizes of 2, 4, 6, 8, 12, 16 or 24, these being the factors of 48. We discuss factors relevant to the choice of block size in the next chapter. At this stage, we simply note that, as a general rule, block size should be roughly equal to the square root of the number of treatments. So, in the above example for $v = 48$, we would consider designs with $k = 6$ or 8. Often site considerations will dictate the block size; physical constraints on the layout of plots in the field might mean that blocks of six plots are more suitable than blocks of eight plots.

An apparent restriction on the availability of alpha designs is that the number of treatments is the product of the number of blocks per replicate and the block size, i.e. $v = sk$. So if we wanted a resolvable incomplete block design for 17 treatments it would appear difficult to achieve. A way around this problem is to allow designs with unequal block sizes. Provided the block sizes differ by no more than one, there is no need to review the model assumption that the pairwise variance between any two plots within a block is the same. Alpha designs make it possible to produce efficient designs with block sizes differing by no more than one. This is done by simply deleting treatment numbers from an alpha design for a larger number of treatments. In Fig. 7.4 we could delete numbers 18, 19 and 20 to obtain a design for 17 treatments in blocks of size three or four. The largest treatment numbers should be deleted before randomisation to guarantee that block sizes differ by no more than one. Alternatively, as we show in Section 7.8, restricted randomisation can be used by specifying appropriate group sizes.

7.5   Extra blocking structures

In the previous section we looked at generalised lattice designs with a nested blocking structure, namely:

$$\text{block} \qquad \text{repl / block / plot}$$

In many situations it is possible to introduce extra blocking structures which serve to provide even more effective control of field variation. In this section we introduce two additional types of blocking structures which can be incorporated in generalised lattice designs:

1. Latinised designs, which provide an extra blocking facility when replicates are contiguous (i.e. next to each other).
2. Resolvable row–column designs, which allow the control of field variation in two directions.

7.5.1   Latinised designs

As mentioned in Section 7.4.3, the layout in Fig. 7.4 does not necessarily reflect the actual configuration of plots in the field. Often, however, we do have a rectangular arrangement of plots for an experiment, a typical case being a glasshouse trial where plots are laid out in a rectangular array. In such situations the replicates are next to each other, or contiguous, and block 1 in replicate 1 might have correspondence to block 1 in the other replicates, i.e. the layout in Fig. 7.4 might be the true plot layout with the replicates contiguous to form a 12×5 array of plots. Then it would not be appropriate to use the randomisation scheme as described in Section 7.4.3, because by chance it might happen that treatment 1 will appear in block 1 in each of the replicates, i.e. in the first (long) column of the rectangular array of plots. While it could be argued that this does not matter because the blocks will make the appropriate adjustments, it would certainly matter if the whole column was seriously damaged. In Chapter 8 we will show that there is also an advantage in being able to remove effects of long columns during the analysis. In any case, it is preferable to include a design feature for such contiguous replicate designs to guarantee that in the long columns of Fig. 7.4 no treatment appears more than once. Designs with this feature are called latinised designs.

By imposing the restriction that no number should appear twice in the column of an alpha array (Section 7.4.2) we can obtain a latinised alpha design. For example, for $r = 4$, $s = 5$ and $k = 4$ we can have the alpha array:

$$
\begin{array}{cccc}
0 & 0 & 0 & 0 \\
1 & 2 & 3 & 4 \\
2 & 4 & 1 & 3 \\
3 & 1 & 4 & 2
\end{array}
$$

leading to the unrandomised latinised alpha design in Fig. 7.5. Note that no treatment number is repeated in any of the five long columns. The strata for a latinised design are *repl*, *column*, *repl.column*, and *repl.column.plot*. In Genstat syntax this would be:

block   (repl * column) / plot

7.5.2 Row–column designs

If the plots of each replicate in Fig. 7.4 are actually laid out in a 4×5 array, it is a good idea to look at a two-dimensional blocking structure, allowing adjustment for field trend using columns, rows or both. This is particularly desirable when plots consist of a square or near-square array of trees. For line plots, the one-dimensional blocking structure described in Section 7.4 is often sufficient.

The unrandomised design in Fig. 7.4 would be very bad as a row–column design, but we have to remember that the randomisation process, particularly the randomisation of plots separately for each block, will result in a much better row

		Block			
1	2	3	4	5	
1	2	3	4	5	
6	7	8	9	10	Replicate 1
11	12	13	14	15	
16	17	18	19	20	
2	3	4	5	1	
8	9	10	6	7	Replicate 2
14	15	11	12	13	
20	16	17	18	19	
3	4	5	1	2	
10	6	7	8	9	Replicate 3
12	13	14	15	11	
19	20	16	17	18	
4	5	1	2	3	
7	8	9	10	6	Replicate 4
15	11	12	13	14	
18	19	20	16	17	

Fig. 7.5. Unrandomised latinised alpha design for r = 4, s = 5 and k = 4.

design, i.e. a design in which rows in Fig. 7.4 are used as incomplete blocks. Nevertheless, the row properties of the design should not be left to chance. Just as we have created an efficient column design in Fig. 7.4, we should also seek an efficient row design and ultimately an efficient row–column design. In Section 7.3 we showed how the column concurrence matrix NN' relates to the average efficiency factor of a one-dimensional blocking structure. Equally, we can define the row concurrence matrix MM' by using the rows of Fig. 7.4 as the incomplete blocks. The average efficiency factor of the row–column design is derived from both NN' and MM'. Details are given by John and Williams (1995, chapter 6).

Construction of efficient row–column designs in general is not easy. The best-known examples are lattice square designs where $k = s$ (Cochran and Cox 1957, chapter 12), but efficient row–column designs can be constructed for other choices of r, s and k using CycDesigN.

The strata for a row–column design are *repl, repl.column, repl.row* and *repl.column.row,* or in Genstat syntax:

$$\text{block} \qquad \text{repl / (column * row)}$$

Researchers Zhou Yunhong and Song Hongliang inspect a second-generation progeny trial of *Eucalyptus pellita* in Chongzuo, Guangxi Province, China. (Photo Roger Arnold)

7.5.3 Latinised row–column designs

The most common layout of plots in glasshouse trials or field trials on favourable sites is plots laid out in a rectangular array. We can then make use of both of the extra blocking structures discussed in this section, namely the latinised property of contiguous replicates and the row–column blocking structure of plots in a rectangular array. In Fig. 7.6 we give an example of a randomised row–column design for $r = 4$, $s = 5$ and $k = 4$ produced by CycDesigN. The average efficiency factor for this design is E = 0.61. Further details on the construction of latinised row–column designs are given in John and Williams (1995, chapter 6).

The strata for a latinised row–column design are *repl, column, repl.column, repl.row* and *repl.column.row*, or in Genstat syntax:

block   (repl / row) * column

The incomplete blocking structures discussed in this section provide a very powerful way of controlling possible field variation, thereby leading to more precise estimation and comparison of treatment means and other factors of interest. This is especially the case when efficient experimental designs are employed, such as those that are generated by CycDesigN.

In general, when the experimental design area is a two-dimensional array of plots, the use of latinised row–column designs is recommended. The design in Fig. 7.6 has the replicates in a line and this is usually the preferred arrangement for

Row	Column					
	1	2	3	4	5	
1	11	8	19	18	12	
2	13	5	20	2	10	
3	1	7	16	14	6	Replicate 1
4	17	3	9	4	15	
1	15	4	3	11	20	
2	5	10	17	9	18	
3	14	16	12	8	13	Replicate 2
4	19	1	2	6	7	
1	18	9	10	1	19	
2	3	2	14	20	8	
3	16	15	7	5	17	Replicate 3
4	6	13	11	12	4	
1	7	17	8	10	2	
2	9	14	1	3	11	
3	12	20	15	19	16	Replicate 4
4	4	6	18	13	5	

Fig. 7.6. Latinised row–column design for $r = 4$, $s = 5$ and $k = 4$.

				Column								
Row	1	2	3	4	5	6	1	2	3	4	5	6
1	3	1	12	11	6	8	23	18	13	14	22	20
2	10	24	9	5	23	18	19	2	8	6	17	4
3	17	13	20	19	15	7	21	1	16	5	12	10
4	14	4	2	22	16	21	3	15	9	7	24	11
1	13	11	15	14	21	12	22	20	7	16	1	3
2	22	6	4	18	9	3	8	24	23	11	14	2
3	1	23	8	20	17	5	10	6	12	18	19	13
4	7	2	19	24	10	16	15	21	4	17	9	5

Replicate 1 (rows 1–4, left), Replicate 2 (rows 1–4, right), Replicate 3 (rows 1–4, left), Replicate 4 (rows 1–4, right)

Fig. 7.7. Resolvable row–column design for r=4, s = 6 and k = 4.

the array of plots. In some cases, however, it is not practical to have replicates in a line. This can occur due to the physical dimensions of the experimental area, or perhaps the need to keep the experiment squarer in nature rather than being extended in one direction. The replicates can then be organised in a two-dimensional arrangement as has been done in Fig. 7.7. This is an example of a latinised row–column design for four replicates, each with four rows and six columns. The replicates are in a 2 × 2 arrangement and assuming they are contiguous in both directions, i.e. the overall experimental area is an 8 × 12 array of plots, then we have the potential to engage latinisation in both the row and column direction. In other words, for this example we can ensure that treatments do not appear more than once in either long rows or long columns.

This two-dimensional arrangement of replicates creates an extra stratum in the latinised row–column analysis of variance table, namely *row*. It should be noted that for smaller designs there is always the danger of introducing too many blocking factors, thereby compromising randomisation. For larger designs, however, it is usually the case that more blocking is an advantage. To that end we have the important theoretical result from Speed *et al.* (1985), namely if in generalised lattice designs the incomplete blocks are not important, we can just revert to a randomised complete block analysis.

Example 7.1

In 2015 a clonal trial of *Eucalyptus* hybrid clones was planted at Ba Vi in Vietnam by the Vietnamese Academy of Forest Science. There were 60 treatments comprising 56 new clones and four standards, one of them a seedling control. CycDesigN was used to construct five replicates of a 10 × 6 row–column design. Each plot was a row of 10 trees at 3 × 2 m spacing. Hence overall there was a 50 × 6 array of plots, i.e. 50 × 60 trees or 150 × 120 m. The trial site was a hill with 10% gradient with replicate 1 at the bottom and replicate 5 at the top.

			Column			
Row	1	2	3	4	5	6
1	21	6	7	41	4	27
2	40	3	38	12	13	44
3	16	57	54	14	19	58
4	11	29	60	51	31	36
5	2	39	46	18	42	15
6	43	33	28	50	52	24
7	30	53	49	59	35	10
8	8	34	56	55	23	45
9	9	5	22	47	20	32
10	17	26	48	1	25	37
1	7	16	37	56	3	50
2	15	59	20	8	24	39
3	36	21	53	46	55	19
4	23	48	2	6	32	51
5	26	31	12	30	54	34
6	28	38	11	44	57	4
7	13	45	41	33	18	22
8	14	10	5	25	60	43
9	29	47	58	35	17	40
10	1	42	52	27	9	49
1	10	18	23	40	33	11
2	50	58	44	60	1	13
3	20	28	31	45	37	6
4	41	8	43	26	38	29
5	34	4	14	24	51	21
6	35	36	3	42	5	54
7	57	17	27	2	22	59
8	32	49	39	7	30	55
9	46	25	47	53	16	52
10	48	12	9	19	15	56
1	45	30	50	29	2	9
2	27	46	55	5	40	31
3	51	19	10	3	28	8
4	60	13	15	49	47	57
5	39	24	16	48	41	35
6	52	32	26	21	44	18
7	25	11	42	58	34	7
8	33	14	17	20	56	38
9	12	22	6	43	36	1
10	54	37	4	23	59	53
1	6	15	40	52	14	3
2	18	20	29	34	27	16
3	24	2	25	38	50	23
4	4	55	33	36	48	30
5	58	51	59	9	26	46
6	5	7	57	10	45	12
7	31	44	19	17	49	41
8	47	54	1	39	11	28
9	22	35	21	37	8	60
10	56	43	32	13	53	42

Replicate 1 (rows 1–10, first block)
Replicate 2 (rows 1–10, second block)
Replicate 3 (rows 1–10, third block)
Replicate 4 (rows 1–10, fourth block)
Replicate 5 (rows 1–10, fifth block)

Fig. 7.8. Field layout for a *Eucalyptus* hybrid clonal trial for r = 5, s = 6 and k = 10.

Looking along a 10-tree row plot of a fast-growing clone within a *Eucalyptus* clone trial at Ba Vi, Vietnam, at age three years. The trial, established by Vietnam's Institute of Forest Tree Improvement and Biotechnology, tests 60 genetic treatments, using six replicates of 10-tree plots (see Examples 7.1 and 8.2).

A suitable specification for this arrangement in CycDesigN would be a latinised design with the treatments in two groups of sizes 4 and 56 with restricted randomisation to ensure the standards (numbers 1…4) are evenly (but not systematically) spread across the rows and columns (Fig. 7.8). This design has an average efficiency factor of 0.743 and the CycDesigN log file is given in Table 7.1. The actual design used for the trial had a slightly lower average efficiency factor of 0.741 but this was still a very good design. An analysis of results from this clone trial is presented in Example 8.2.

Table 7.1 Log file from CycDesigN for construction and randomisation of the Fig. 7.8 layout.

Date and time: Mon Aug 29 07:48:23 2022

Design properties
    Resolvable

Two stage
    Row–column design
    Single factor

Design parameters

Number of treatments	=	60
Number of rows	=	10
Number of columns	=	6
Number of replicates	=	5

Latinized by columns

Nested (Optimal)
Number of nested groups = 24

Group 1	Size 4	Treatments 1, 2, ...
Group 2	Size 56	Treatments 5, 6, ...

Random number seed for design generation = 602

Average efficiency factors		(Upper bounds)	(Weights)
Row	0.824964	(0.829234)	(0.079)
Column	0.899764	(0.900432)	(0.132)
Row–column	0.743493	(0.758936)	(0.789)

Concurrencies

Concurrence	Row	Column
0	1039	708
1	712	801
2	19	234
3	0	27

Random number seed for design randomization = 457

7.6 Spatial enhancements

7.6.1 Resolvable designs

The randomisation of the blocking factors for the latinised row–column design in Fig. 7.7 would involve randomising the order of long columns in replicates 1/3 and 2/4 and randomising the order of long rows in replicates 1/2 and 3/4. And whilst this randomisation and subsequent design layout would possibly satisfy a theoretical statistician, users can often comment when treatment patterns occur by chance. For example in Fig. 7.7, treatment 21 has replicates diagonally adjacent to each other, as does treatment 7. These are examples of treatment clumping and it is something that the use of resolvable designs helps to minimise. Hence in this instance the issue is not too severe; nevertheless users often feel more comfortable if the layout can avoid any such juxtapositions. Even the existence of knight's move proximities can raise eyebrows; for example treatment 23 in replicates 1 and 2, treatment 4 in replicates 1 and 3 and treatment 16 in replicates 2 and 4.

CycDesigN has facilities to not only generate latinised row–column designs with replicates in two-dimensional arrangements, but also to ensure that the occurrence of diagonal and knight's move self-adjacencies are minimised, e.g. see Fig. 7.9. Another aspect of the row and column randomisation is the possibility that two treatments will be in neighbouring plots more than once. Again this can be viewed as a negative by users and probably for good reason. For example if one of the treatments fails in all replicates then the neighbouring plots could attain an advantage. Alternatively if the treatment substantially outperforms others then the neighbours could be disadvantaged. If plots are long and narrow, e.g. line plots in tree breeding or as they typically are in cereal breeding, then the possible concern might only be in one direction, i.e. the long boundaries. Looking at Fig. 7.7 there are a number of instances where the same two treatments are neighbours twice, e.g. treatments 1 and 3 are row neighbours in replicates 1 and 4. CycDesigN has a spatial option to address multiple occurrences of treatments as neighbours; the package tries to minimise a quantity called neighbour balance (NB). If two treatments are neighbours more than once in rows then this contributes one to the row NB. There is a similar definition for column NB. So in Fig. 7.9 there are no instances of pairs of treatments as neighbours more than once in either the row direction or the column direction; i.e. we have an NB score of 0 in both directions.

Needless to say, nothing comes for free and so there can be a price to pay for these spatial considerations. This can come in the form of:

	Row	1	2	3	4	5	6	1	2	3	4	5	6	
							Column							
Replicate 1	1	1	14	15	21	8	2	18	16	20	19	17	11	Replicate 2
	2	23	6	13	12	22	17	3	10	15	2	9	7	
	3	3	20	24	10	11	19	4	8	5	23	14	12	
	4	16	7	4	18	9	5	21	6	22	24	13	1	
Replicate 3	1	21	13	22	16	2	7	1	9	17	10	6	18	Replicate 4
	2	5	18	1	24	12	6	19	23	7	15	4	22	
	3	11	23	10	14	17	15	13	20	8	12	2	3	
	4	19	8	9	20	4	3	16	21	24	5	11	14	

Fig. 7.9. Resolvable spatial design for r=4, s = 6 and k = 4.

1. A possible drop in the average efficiency factor (E); this is usually small and for larger designs there is often no change at all. For the designs in Figs 7.7 and 7.9 the average efficiency factor drops from 0.63 to 0.61.
2. A restriction on the randomisation appropriate for the blocking structure of the experimental design. Then there is the possibility of creating a bias on the significance levels that we use for testing. For example in the analysis of variance table we might carry out an F-test at the nominal 5% level to evaluate the significance of a factor of interest. This means that a significant result might happen by chance once in every 20 times. A bias can mean that this percentage level will vary. It has been shown, however, that any bias is very small. More detail can be obtained in Williams and Piepho (2019) and the references therein.

7.6.2 Non-resolvable designs

In general resolvable designs are to be recommended wherever possible. There are, however, some situations that would make their use difficult to implement. These include:

1. Due to say, lack of resources, the required number of replications cannot be made for one or more treatments. Whilst it would be undesirable to have replication numbers varying greatly across treatments, differences of just one can sometimes occur.
2. There are restrictions on the experimental site that preclude a suitable structure for a resolvable design. For example the plots might be in a 7×15 array and we want a design for 26 treatments. Then it would be better to consider a non-resolvable design with one of the treatments replicated five times and the remainder replicated four times.

But in using a non-resolvable design we are much more likely to run the risk that after designing the experiment to optimise the allocation of treatments to rows and columns, and then randomising the order of both of these blocking factors, the replications of one or more treatments might clump together. One form of clumping can be seen occurring for treatment 15 in Fig. 7.10 where the four replicates span just five columns of the design. Another form, as mentioned in the previous section, is where replications of the same treatment appear as diagonal self-adjacencies, or in a knight's move pattern. Users do not like any forms of clumping of treatment replications and will often carry out new randomisations of rows and columns to try to minimise the clumping effect. We can, however, restrict the design construction to avoid clumping as much as possible whilst suffering little or no reduction in the average efficiency factor of the design.

							Column								
Row	1	2	3	4	5	6	7	8	9	10	11	12	13	14	15
1	24	7	4	3	5	23	6	21	18	15	25	2	19	16	8
2	6	1	11	12	22	19	18	15	7	3	20	9	8	14	13
3	10	17	20	25	19	5	2	9	26	22	23	1	12	6	14
4	23	8	6	18	4	9	15	1	14	26	3	24	21	17	25
5	13	21	25	19	24	16	17	10	22	1	9	11	26	7	4
6	8	20	26	13	18	15	10	12	2	16	7	5	11	24	9
7	22	5	16	17	20	14	11	4	21	13	10	3	23	12	2

Fig. 7.10. Non-resolvable row–column design for v=26, s = 15 and k = 7.

							Column								
Row	1	2	3	4	5	6	7	8	9	10	11	12	13	14	15
1	15	24	3	9	13	8	18	10	21	5	17	11	25	2	16
2	21	6	4	25	26	22	23	2	20	1	16	14	7	18	24
3	19	8	16	17	1	15	14	11	4	13	23	12	22	10	5
4	13	5	7	21	12	3	25	6	17	18	15	20	26	1	9
5	12	26	23	10	11	20	24	7	16	19	2	6	8	17	3
6	11	1	18	14	6	19	26	15	8	22	24	9	21	4	23
7	25	20	22	2	4	16	5	9	12	10	7	3	13	19	14

Fig. 7.11. Non-resolvable spatial design for v=26, s = 15 and k = 7.

The spatial features in CycDesigN do this by ensuring that the replications of every treatment span as many rows and columns as possible. So in Fig. 7.11 the replications of all treatments span at least 11 columns and five rows. We can compare this with Fig. 7.10 where the minimum spans are five and four for columns and rows respectively. The drop in average efficiency factor in achieving this much more desirable spread of treatment replications is minimal, from 0.85 to 0.83. These spatial features, namely treatment spans, diagonal self-adjacencies, knight's move adjacencies and several other less-severe adjacencies are called evenness of distribution (ED) of treatment replications.

The spatial design in Fig. 7.11 also addresses the problem of neighbour balance as evident in multiple instances of pairs of treatments appearing together twice in rows in Fig. 7.10; in fact treatments 15 and 18 occur as row neighbours three times. On the other hand Fig. 7.11 has a row NB score of zero. Note that for an arrangement like this, with many more columns than rows, we would expect that the plots would be long and narrow, thereby making the experimental area squarer in nature. Hence the neighbours in the row direction are likely to be more important than those in the column direction and so for non-resolvable designs, the neighbour balance score is only considered in the row direction.

A detailed discussion of NB&ED designs is given by Piepho *et al.* (2021).

Mick Underdown inspects a young provenance-progeny trial of spotted gum (*Corymbia maculata*) planted near Hamilton in western Victoria, Australia. The trial tested 108 families and used a resolvable incomplete block design with nine blocks of 12 five-tree plots per replicate. The young trees were protected from browsing by tree guards.

7.7 Partially replicated designs

In most ongoing breeding programs the screening of large numbers of candidates (progenies or clones) is undertaken. There is a wide variety of approaches adopted to carry out the testing and selection of material for further research. Rezende *et al.* (2014) discuss stage 1 *Eucalyptus* clone trials where a large number of candidate clones are evaluated across a range of sites using single-tree plots with a small number of replications per site. Son *et al.* (2018) describe stage 1 trials of about 200 candidate *Acacia* hybrid clones using five or more complete replicates and single-tree or two-tree plots, with commercial clones included as standards.

Federer (1956) introduced augmented designs using replicated standards and unreplicated entries within a site. In the analysis the replicated standards allow the estimation of an error term for the comparison of the entries. In order to adequately estimate this error 15–20% of the plots are usually assigned to standards. More recently Cullis *et al.* (2006) introduced the idea of partially replicated (p-rep) designs. Here a percentage of the entries for evaluation are replicated at a site whilst the remainder are unreplicated. These designs are likely to be more efficient than using augmented designs, mainly because the number of plots allocated to standards can be reduced whilst still maintaining the same level of precision for testing. It is, however, still desirable to include one or more replicated standards in a p-rep design to allow a comparison of new entries with existing material.

P-rep designs can be implemented at a single site or across multiple sites (Möhring *et al.* 2014); CycDesigN can construct efficient designs for both these situations and for the latter in particular, the package ensures that the overall design is connected (John and Williams 1995, section 1.8). Furthermore the spatial enhancements discussed in the last section can be invoked to address the row and column spans of the standards and the duplicated entries. As an example Fig. 7.12 is a p-rep row–column design in a 10 × 15 array and for 106 entries, 34 of which have been duplicated and these have been printed in bold. Also included is a single standard treatment, replicated 10 times and spread evenly (but not systematically) across rows and columns. Hence there is a

	Column														
Row	1	2	3	4	5	6	7	8	9	10	11	12	13	14	15
1	23	**68**	**56**	**52**	59	84	S	**3**	33	**63**	**89**	73	47	**45**	**90**
2	**17**	**18**	83	**51**	46	**11**	7	4	**21**	93	32	43	78	**42**	S
3	S	9	24	28	25	**1**	**30**	**37**	86	**76**	81	106	**104**	35	**44**
4	20	**61**	**75**	92	S	100	**45**	8	60	69	**85**	87	**40**	**67**	99
5	**21**	15	**57**	65	**50**	36	**26**	97	S	**56**	58	5	**10**	72	19
6	**76**	S	**62**	101	71	14	**90**	34	27	**68**	22	**54**	**39**	**11**	29
7	98	**85**	**13**	**89**	**67**	16	70	103	48	2	S	**66**	94	**30**	**51**
8	**37**	88	49	96	**40**	31	41	S	**17**	74	**75**	**50**	55	12	**3**
9	**104**	**42**	6	**39**	95	S	102	53	**61**	**57**	79	**18**	**62**	**52**	**26**
10	105	**63**	**54**	82	**66**	**10**	64	**44**	80	38	77	S	**1**	91	**13**

Red: Standard **Bold**: Duplicate

Fig. 7.12. Spatial p-rep row–column design for 106 entries with a replicated standard.

standard in each row of the design and at most one standard in each column. Furthermore, the spatial enhancements have ensured that the replicated entries are well spread out with a separation of at least eight columns and four rows. For example, entry 89 is in columns 4 and 11 and entry 85 is in rows 4 and 7; all other replicated entries have this or greater row and column separations. The average efficiency factor for this design is 0.59. This example, however, is just for illustration; typically p-rep designs would be employed for very many more entries per site and for multiple sites.

7.8 Using CycDesigN

The computer software package CycDesigN (VSN International) provides a convenient means of constructing alpha, row–column, latinised, spatial and partially replicated designs for a wide range of useful parameter combinations. Some examples of the designs available are given by Williams *et al.* (1999) and Piepho *et al.* (2021). The extensive online help facilities describe the package in detail; here we will simply outline the structure and components discussed in this book:

1. CycDesigN is a Windows-based package. The user can specify whether a block design or row–column design is required. There is an option to construct resolvable or non-resolvable designs; for tree improvement trials we would normally choose resolvable designs (Section 7.4.1).
2. The basic parameters of the design are entered, namely r, v, s and k. CycDesigN can generate designs for a single treatment factor or factorial designs (Section 4.3). Treatment factors can also be nested, e.g. seedlots within provenances (Section 8.4).
3. CycDesigN can produce different types of latinised designs to accommodate the arrangement of replicates in the field. It is normally preferable to have replicates contiguous in a line, as in Fig. 7.6. Sometimes, however, for a design with many contiguous replicates, a two-dimensional arrangement of replicates is sensible. CycDesigN can provide the latinised feature in two directions.
4. The discussion of latinised designs in Section 7.5.1 only considered the feature as applied to a single long column. CycDesigN, however, extends to the construction of t-latinised designs (John and Williams 1998) where the latinised feature is applied to sets of t columns. In this notation the designs of Section 7.5.1 would be 1-latinised.
5. Spatial designs (as add-ons to block and row–column designs) with good neighbour balance and evenness of distribution (NB&ED) properties can be generated to avoid the occurrence of undesirable treatment patterns and clumping after randomisation.

6. CycDesigN can generate efficient partially replicated (p-rep) designs for single or multiple sites. Users can specify the number of entries, trial dimensions and whether standards are to be incorporated.

7. Once a design has been constructed, CycDesigN uses the randomisation procedure appropriate for the design. This will vary depending on which design features have been incorporated, e.g. block or row–column design, type of latinisation, nested treatment structure etc. The user specifies the randomisation 'seed'; hence different randomisations can be produced from the same design construction, as might be required for the same experiment at several sites.

8. CycDesigN can allow researchers to use numerical treatment labels in the final design layout. As an example, seedlots are numbered in the nursery but not all seedlots germinate or can be raised in sufficient quantity for field planting. Therefore, the design required is for a smaller number of seedlots than we started with, but at the same time we want to retain original seedlot numbers. This can be done in CycDesigN.

9. The record of the CycDesigN session is saved in a log file so that the user has details of options selected and randomisation carried out to produce a final layout. The package also generates a spreadsheet file of design factors ready for data entry and analysis by standard statistical packages.

8

Analysis of generalised lattice designs

8.1 Introduction

In Chapter 7 we introduced the concept of incomplete block designs and defined the class of generalised lattice designs. In experiments with large numbers of treatments, the advantages of incomplete blocking structures may be considerable. With these advantages, however, comes a more complex construction and layout. Nevertheless, any difficulty is largely overcome by a package such as CycDesigN. At the other end of the process we also need an effective means to analyse generalised lattice designs and thus realise the gains that they offer. While many smaller statistical packages cannot undertake the analysis of generalised lattice designs, most powerful packages provide a convenient facility for doing so.

The theory on which parameter estimation for generalised lattice designs is based involves a lot of matrix algebra, so we will not go into detail. Instead, we will use worked examples to illustrate how to carry out analyses. In Section 8.2 we will carry out the analysis of linear models of the type presented in Chapters 4 and 7, known as fixed-effects models. For many non-orthogonal designs, however, the use of fixed-effects models results in a less efficient analysis and is not recommended. Hence, in Section 8.3 we present a more appropriate model, known as a mixed-effects model, for the analysis of generalised lattice designs. In Genstat the **fit** command (introduced in Chapter 5) is used for the fixed-effects analysis of variance of a non-orthogonal design, while the **reml** command (introduced in Chapter 6) can be used for both fixed- and mixed-effects models.

8.2 Fixed-effects model

The model for a generalised lattice design with just one incomplete blocking structure was given in Section 7.3 as:

$$Y = G\mu + R\rho + Z\beta + X\tau + \varepsilon \tag{8.1}$$

This is known as a fixed-effects model because the overall mean, replicate, block and treatment parameters are assumed to be fixed and require estimation, compared with the vector of residuals ε, the elements of which are assumed to be random with an underlying normal distribution.

An advantage of incomplete blocking structures is that they provide a mechanism to better account for site variation. It does not follow, however, that we gain most by making the blocks as small as possible. The reason for this is that as blocks become smaller, fewer treatments will be represented in a block and hence we can make fewer precise comparisons between treatments. We will use a simple example to explain this further.

Consider an RCB design with two replicates and four treatments:

	Replicate	
	1	2
	1	1
	2	2
	3	3
	4	4

$$\tag{8.2}$$

In replicate 1 we can make pairwise comparisons between treatment parameters τ_1 and τ_2, τ_1 and τ_3 etc. – six comparisons in all. Thus within the two replicates a total of 12 pairwise comparisons can be made between treatments. On the other hand, in the following generalised lattice design for $r = s = k = 2$:

		Replicate			
		1		2	
Block		1	2	1	2
		1	3	1	2
		2	4	3	4

$$\tag{8.3}$$

only four pairwise comparisons can be made within the blocks. There are, however, other indirect comparisons that can be made within blocks. For example, treatment parameters τ_1 and τ_4 can be compared indirectly by making the pairwise comparisons $\tau_1 - \tau_2$ in block 1 of replicate 1 and $\tau_4 - \tau_2$ in block 2 of replicate 2. The difference $(\tau_1 - \tau_2) - (\tau_4 - \tau_2) = (\tau_1 - \tau_4)$ provides information on the comparison of treatment parameters τ_1 and τ_4. This is still a within-block treatment comparison.

There are other treatment comparisons associated with block totals. For example, the difference between the totals in block 2 of replicates 1 and 2 would not only involve the block parameters β_{12} and β_{22}, but also the comparison $(\tau_3 + \tau_4) - (\tau_2 + \tau_4) = (\tau_3 - \tau_2)$ of treatment parameters. In summary, what has happened is that because the design (8.3) is non-orthogonal, the information available on treatment differences has been split. Some information is contained in within-block comparisons, while the remainder is tied up in between-block totals. In fact, the proportion of treatment information available from within-block comparisons is given by the average efficiency factor (E) for the design (see Section 7.3). For the design in (8.3), $E = 0.6$, which means that only 60% of the treatment information is contained in within-block comparisons. As the block size gets smaller, E and hence the amount of within-block treatment information decreases.

A problem with the fixed-effects model (8.1) is that it uses only within-block treatment information for the estimation of treatment parameters, so the analysis is also known as an intra-block analysis. The RCB design in (8.2) has all the treatment information available within replicates ($E = 1.0$), so in using (8.1) we need to know whether introducing the incomplete blocking structure has led to a reduction in the magnitude of the residual mean square that more than compensates for the loss of treatment information. Whether or not this happens is a function of the block size (k) and the magnitude of the site variation. If k is too small then E will be low and the reduction in residual mean square required to compensate for this will be too great. Alternatively, if there is very little site variation, the incomplete blocking structure again will not sufficiently reduce the residual mean square. In Section 8.3 we introduce a different model which is appropriate in many situations and avoids some difficulties associated with the fixed-effects model. But for the remainder of this section we will give an example of the fixed-effects analysis.

Example 8.1

In the early 1990s Khongsak Pinyopusarerk of CSIRO Forestry and Forest Products initiated a far-reaching study of *Casuarina equisetifolia*. This is a nitrogen-fixing tree of considerable social, economic and environmental importance in tropical/subtropical littoral zones of Asia, the Pacific and Africa. Provenance collections were obtained from 18 countries and, with this material, more than 40 trials were laid out in 20 countries. The number of seedlots included in each trial varied, depending on the suitability and size of the planting sites for the available material. One of the trials, in Weipa, northern Queensland, contained all the available seedlots and is the example used here. Further details are given in Pinyopusarerk and Williams (2000).

8.2.1 Experimental design

There were originally 64 seedlots available for germination in the nursery. Four seedlots (Nos. 9, 43, 44 and 49) failed to germinate, leaving 60 seedlots for the trial. A latinised row–column design for four replicates, each with six rows and 10 columns, was designed. Each plot consisted of a 5 × 5 arrangement of trees at 2 × 2 m spacing. The full layout of the trial is given in Fig. 8.1 and corresponds to an array of 120 × 50 trees with dimensions 240 × 100 m. Note that:

1. The design of this trial was carried out before the package CycDesigN became available. The average efficiency factor for the layout in Fig. 8.1 can be improved with the use of CycDesigN.
2. The layout in Fig. 8.1 uses the seedlot numbers as defined in the nursery, i.e. 1 to 64, with the four non-germinating seedlots missing. This is the recommended procedure (Section 2.4) and CycDesigN has facilities to produce a layout in this form.
3. Unfortunately in this trial, seedlot 64 was not planted and so there were actually only 59 provenances laid out in the field.

| | | | | | Column | | | | | | |
Row	1	2	3	4	5	6	7	8	9	10	
1	20	51	8	27	13	56	16	25	54	45	
2	36	11	32	38	2	37	29	48	17	23	
3	21	41	58	52	15	60	26	4	30	14	Replicate 1
4	24	64	35	53	22	61	10	6	42	12	
5	59	55	18	28	31	33	40	62	63	46	
6	34	3	50	19	7	57	47	39	5	1	
1	4	63	31	40	36	20	14	47	25	6	
2	19	15	24	54	28	30	55	3	41	61	
3	29	57	42	18	32	26	17	59	58	34	Replicate 2
4	45	53	16	23	56	2	22	60	7	13	
5	5	21	46	51	1	38	48	52	12	10	
6	50	39	11	37	62	35	27	8	33	64	
1	31	13	20	7	37	51	8	56	26	16	
2	38	17	28	48	29	59	1	64	3	5	
3	58	35	33	21	12	27	42	54	14	32	Replicate 3
4	62	23	63	50	30	15	57	22	36	2	
5	55	47	40	10	11	46	19	45	18	41	
6	61	4	6	60	39	25	34	24	53	52	
1	48	56	2	42	26	23	20	12	32	55	
2	40	16	29	62	58	63	61	5	13	47	
3	15	8	34	22	6	36	4	10	19	39	Replicate 4
4	52	30	25	33	17	64	41	38	46	24	
5	11	27	14	3	45	31	54	57	21	18	
6	1	7	60	35	59	50	53	51	37	28	

Fig. 8.1. Field layout of plots for *Casuarina equisetifolia* provenance trial.

Table 8.1   Country of origin of the provenances in Example 8.1.

Country Number	Name	Provenance Number
1	Australia	1, 2, 3, 4
2	Benin	5
3	China	6, 7, 63
4	Egypt	8, 10, 11
5	Fiji	12, 13, 56
6	Guam	14
7	India	15, 16, 17, 18, 19, 20
8	Kenya	21, 22, 23, 24, 25, 26, 27, 28
9	Malaysia	29, 30, 31, 32, 33, 34, 35, 36, 57
10	Mauritius	58
11	PNG	37
12	Philippines	38, 39, 40
13	Puerto Rico	59
14	Solomon Is.	41, 42
15	Sri Lanka	60, 61, 62
16	Thailand	45, 46, 47, 48
17	Vanuatu	50, 64
18	Vietnam	51, 52, 53, 54, 55

4. The latinised row–column design employed for this trial meant that individual seedlots were represented no more than once in the plots in each long column. Although there seemed no prior reason for introducing the latinised blocking structure, it turned out to be a critical design feature, as discussed in Section 8.3.

5. The 59 seedlots used in the trial were grouped by country of origin and this nested structure was incorporated into the design as discussed in Section 7.4. Details of this structure are given in Table 8.1.

6. To increase the production of nitrogen nodules, the seedlings were inoculated with *Frankia* culture isolated from nodules of *Casuarina equisetifolia*. Because available inoculum was not sufficient for all seedlings in one application, half the seedlings were inoculated seven weeks before planting and the other half shortly before planting. The seedlings from the longer inoculation time were used in replicates 1 and 4 (as a result of randomisation) and the other seedlings for replicates 2 and 3. The idea of confounding treatment factors with replicates has been mentioned in Section 2.5.

8.2.2   RCB analysis

Several variates were measured at 30 months including the quantitative variates height, diameter at breast height in cm (*dbh*) and a number of qualitative variates; here we will concentrate on *dbh*. A plot summary file was created and the plot means are given in Table B8.1. Note that some of the trees had more than one stem, in which case the *dbh* of each stem was measured and the *dbh* of a single stem with the same basal area was derived. For the small number of trees with more than

Casuarina equisetifolia provenance trial at Weipa, Queensland (aged four months), as discussed in Example 8.1. Already the different form of the provenances is evident. (Photo Khongsak Pinyopusarerk)

three stems, only the three largest were measured. The plot variances were screened and an initial RCB analysis was carried out. The Genstat program is given in Table C8.1 and the output in Table 8.2. The plot of residuals against fitted values (Fig. 8.2) indicates that the data are well-behaved. The RCB analysis gives a residual mean square of 0.6057 and there are significant differences between countries (P < 0.001).

8.2.3　Fit analysis

The latinised row–column fixed-effects model is an extension of (8.1) and includes terms for rows and columns within replicates as well as a term for the long columns. The non-orthogonal analysis of variance uses the **fit** command in Genstat (Appendix A). The user simply specifies the factors which are to be included in the model. The Genstat program and output are given in Tables C8.2 and 8.3 respectively. It can be seen that in the fixed-effects model the long columns as well as the rows and columns within replicates are significant, so this is a good indication that it is worth proceeding to the more appropriate mixed-model analysis as discussed in the next section. If any of the incomplete blocking structures had not been significant they could have been left out of

Table 8.2 Part of the Genstat output from the RCB analysis of Example 8.1.

Analysis of variance

Variance: v[1]; dbh – dbh=sqrt(dbh1*dbh1 + dbh2*dbh2 + dbh3*dbh3)

Source of variation	d.f.	s.s.	m.s.	v.r.	F pr.
repl stratum					
inoc	1	11.5415	11.5415	11.46	0.077
Residual	2	2.0142	1.0071	1.66	
repl.row.column stratum					
country	17	54.6185	3.2129	5.30	<.001
inoc.country	17	10.0724	0.5925	0.98	0.487
country.prov	41	18.6057	0.4538	0.75	0.854
inoc.country.prov	41	21.4625	0.5235	0.86	0.698
Residual	116	70.2557	0.6057		
Total	235	188.5705			

Tables of means

Variance: v[1]; dbh – dbh=sqrt(dbh1*dbh1 + dbh2*dbh2 + dbh3*dbh3)
Grand mean 3.404

inoc	1 week	7 weeks		
	3.183	3.625		

country	Australia	Benin	China	Egypt	Fiji
	2.631	3.343	3.686	2.498	2.612
rep.	16	4	12	12	12

country	Guam	India	Kenya	Malaysia	Mauritius
	2.343	3.575	3.491	4.031	3.123
rep.	4	24	32	36	4

country	Philippines	PNG	Puerto Rico	Solomon Is.	Sri Lanka
	3.612	3.650	3.345	3.699	3.243
rep.	12	4	4	8	12

country	Thailand	Vanuatu	Vietnam
	3.841	2.763	3.276
rep.	16	4	20

Standard errors of differences of means

Table	inoc	country	inoc country	country prov	
rep.	118	unequal	unequal	4	
s.e.d.		0.5503	0.7826		min.rep
d.f.		116	117.16		
s.e.d.	0.1306	0.4102	0.5859	0.5503	max-min
d.f.	2	116	110.85	116	
s.e.d.		0.1834X	0.2722		max.rep
d.f.		116	31.61		

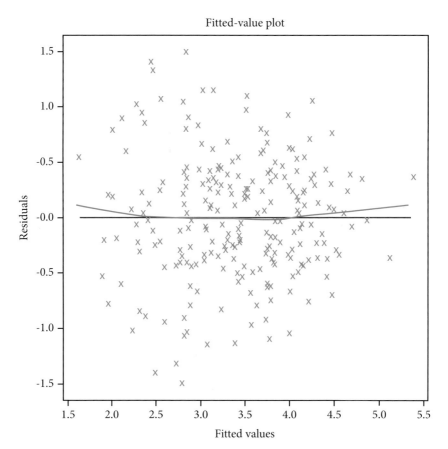

Fig. 8.2. Plot of residuals against fitted values from RCB analysis of variance of *dbh* means for Example 8.1.

the model, resulting in a simpler analysis, ultimately reducing to an RCB analysis.

Note that the **fit** command in Genstat does not separate the analysis into strata, as happens with **anova**. Therefore care is needed to ensure that the correct error term is used for calculation of variance ratios. The residual mean square in Table 8.3 is 0.2402 (down from 0.6057 for the RCB analysis). This results from the fact that the incomplete blocking factors, namely long columns and rows and columns within replicates, are all highly significant, i.e. they are being effective in controlling field variation. The provenance differences are also highly significant but, at this stage of the non-orthogonal analysis, we have not partitioned the provenances into countries and provenances within countries (as was done in Table 8.2). We prefer to do this as a separate exercise (discussed in Section 8.4). Note also that the interaction between inoculation and provenances is not significant and we may prefer to drop this from the model. Since the mean square

---

Table 8.3   Part of the Genstat output from the fixed-effects analysis of Example 8.1.

Regression analysis

Response variate: v[1]; dbh – dbh=sqrt(dbh1*dbh1 + dbh2*dbh2 + dbh3*dbh3)
Fitted terms: Constant + inoc + repl + column + repl.row + repl.column + prov + inoc.prov

Accumulated analysis of variance

Change	d.f.	s.s.	m.s.	v.r.	F pr.
+ inoc	1	11.5415	11.5415	48.05	<.001
+ repl	2	2.0142	1.0071	4.19	0.020
+ column	9	65.2421	7.2491	30.18	<.001
+ repl.row	20	16.5932	0.8297	3.45	<.001
+ repl.column	27	16.4051	0.6076	2.53	0.001
+ prov	58	53.8939	0.9292	3.87	<.001
+ inoc.prov	58	8.4698	0.1460	0.61	0.971
Residual	60	14.4107	0.2402		
Total	235	188.5705	0.8024		

---

for the interaction is less than the residual mean square, dropping the term will reduce the value of the residual mean square to 0.1939.

8.3   Mixed-effects model

In the previous section we discussed a problem of the fixed-effects model for non-orthogonal designs, namely that we sacrifice some treatment information so that we can estimate parameters to control site variation. But if gain in site control does not offset the loss of treatment information, we can be worse off than simply doing an RCB analysis. In many experimental situations the incomplete blocking structures we impose are merely groupings of plots rather than representing a distinct entity, such as a constant temperature cabinet, and thus we can never be sure of the degree to which a blocking structure will improve the analysis. We therefore want a model that does not force us to estimate block parameters if they are not needed. Such a model is provided by the mixed-effects model, in which blocks are assumed to be random effects. The model can be written as:

$$Y = G\mu + R\rho + X\tau + Z\xi + \varepsilon \tag{8.4}$$

where ξ is a vector of random effects from a normal distribution with zero mean, and variance σ_b^2.

Although (8.4) looks similar to (8.1), the differences are considerable. Instead of having to estimate a set of block parameters β, now we only have to estimate the block variance component σ_b^2. An alternative formulation of (8.4) makes this clearer. Since the random effects are all assumed to be independent we can write (8.4) as:

$$Y = G\mu + R\rho + X\tau + \eta \tag{8.5}$$

where $\mathrm{Var}(\eta) = \sigma^2 I + \sigma_b^2 ZZ'$

An implication of this model is that the pairwise variance between two plots in the same block is $2\sigma^2$ whereas that between two plots in different blocks of the same replicate is $2(\sigma^2 + \sigma_b^2)$.

Several methods are available for parameter estimation in mixed-effects models. The most popular is residual maximum likelihood (REML), where the variance components σ^2 and σ_b^2 are estimated by iteration then substituted in the expression for $\mathrm{Var}(\eta)$. Then (8.5) is solved using generalised least squares (Patterson and Thompson 1971). The variance components can also be estimated by equating sums of squares in an analysis of variance table to expectation, as done by Cochran and Cox (1957, chapter 10). This method was popular a few years ago before the more powerful computers required for the preferred REML estimation became common.

The major advantage of the mixed-effects model is that estimation of treatment means uses all treatment information, rather than just the within-block information used in the fixed-effects model. What happens is that the proportion E (the average efficiency factor, see Section 7.3) of treatment information available within blocks is assigned a weight $1/\sigma^2$ and between-block treatment information is given a lower weight, $1/(\sigma^2 + \sigma_b^2)$. Now we can look at two extremes:

1. When the blocks are not helping to reduce site variation, i.e. $\sigma_b^2 = 0$, the within- and between-block treatment information is combined with the same weight and the resulting estimated means are identical to those from an RCB analysis.

2. When there is large site variation within replicates and blocks have been able to effectively adjust for this variation, σ_b^2 will be very large so the weight given to the between-block treatment information will be near zero. Then the estimated treatment means will be the same as those obtained from the fixed-effects model.

In practice, the estimation of treatment means will lie somewhere between the above extremes, but the value of the mixed-effects model is that the procedure will react to the contribution provided by the blocks – if this contribution turns out to be zero, the analysis will revert to the RCB analysis. Hence this type of analysis (also known as recovery of inter-block information) is never worse than RCB analysis, but if the blocks are helping then the analysis will be much better than RCB analysis in terms of increasing precision for the comparison of estimated treatment means. The use of generalised lattice designs rather than RCB designs can therefore be viewed as a type of free insurance. When the blocks are helping we gain; when they are not needed (i.e. non-significant), they can be dropped from the analysis and we are no worse off. It follows that experimenters should use generalised lattice designs wherever possible and, if appropriate, more than one incomplete blocking structure should be employed.

Latinised designs may have considerable advantage when it comes to recovery of inter-block information. This is because in many cases some variation due to site can be incorporated in long-column differences which are nearly orthogonal to treatments. Then the estimate for the column-within-replicate variance component will be smaller than if long columns had not been used, hence the weighting for between columns in the recovery of treatment information will be greater, thereby increasing overall precision for comparisons of estimated treatment means (Williams 1986). Note that we normally specify the long columns in latinised designs as fixed; this is because the long column effects are often large, thereby reducing the prospect of recovery of treatment information. Also, the long columns may represent a particular feature and so the fixed-effects specification would be more appropriate.

Example 8.1 (continued)

The non-orthogonal analysis of variance in Table 8.3 showed that the incomplete blocking factors (long columns and rows and columns within replicates) were effective in reducing the residual mean square compared with an RCB analysis. Having established a need for the incomplete blocking factors, we can now use the mixed-model formulation to enable the recovery of provenance information from between rows and columns within replicates. This is done by specifying these factors as random in a straightforward extension of (8.5). The Table 8.3 analysis also showed that the interaction between inoculation and provenances was not significant, so it is reasonable to drop this from the mixed-effects formulation. Although it could have easily been left in, we have also dropped the inoculation main effect from the model since it is totally confounded with replicates and we have already obtained the requisite information on the effects of inoculation time from Table 8.2.

The Genstat program for the REML analysis of the mixed-effects model is given in Table C8.3, with output shown in Table 8.4. The estimated variance component for rows within replicates (*repl.row*) is 0.0640 with a standard error of 0.0297. We have already observed from the fixed-effects non-orthogonal analysis of variance in Table 8.3 that rows within replicates are significant; this is supported in the mixed-model analysis where the variance component is more than twice the standard error. A similar but less strong result exists with the variance component for columns within replicates (*repl.column*), but it is worth remembering that this variance component measures variation after the long columns have been fitted.

The estimate for σ^2 is 0.195, which can be compared with the value of 0.6057 for the RCB model. In the Table C8.3 program we print out both the mixed model and RCB estimated provenance means. Further analyses can be carried out on these means (discussed in Section 8.4).

Table 8.4   Genstat output from the mixed-effects analysis of Example 8.1.

REML variance components analysis

Response variate: v[1]; dbh – dbh=sqrt(dbh1*dbh1 + dbh2*dbh2 + dbh3*dbh3)
Fixed model: Constant + repl + column + prov
Random model: repl.row + repl.column
Number of units: 236

Residual term has been added to model

Sparse algorithm with AI optimisation

Estimated variance components

Random term	component	s.e.
repl.row	0.0640	0.0297
repl.column	0.0459	0.0260

Residual variance model

Term	Model(order)	Parameter	Estimate	s.e.
Residual	Identity	Sigma2	0.195	0.0254

Tests for fixed effects

Sequentially adding terms to fixed model

Fixed term	Wald statistic	n.d.f.	F statistic	d.d.f.	F pr.
repl	12.82	3	4.27	21.9	0.016
column	144.93	9	16.10	23.3	<0.001
prov	284.06	58	4.90	133.3	<0.001

Dropping individual terms from full fixed model

Fixed term	Wald statistic	n.d.f.	F statistic	d.d.f.	F pr.
repl	13.22	3	4.41	21.9	0.014
column	127.46	9	14.16	23.3	<0.001
prov	284.06	58	4.90	133.3	<0.001

Message: denominator degrees of freedom for approximate F-tests are calculated using algebraic derivatives ignoring fixed/boundary/singular variance parameters.

Table of predicted means for repl

rep	1	2	3	4

	3.754	3.127	3.213	3.502

Standard errors of differences

Average: 0.1927
Maximum: 0.1927
Minimum: 0.1927

Table of predicted means for column

column	1	2	3	4	5	6	7	8
	3.505	3.500	3.851	3.828	3.595	3.783	3.306	3.516

column	9	10
	3.178	1.930

Standard errors of differences
Average: 0.2086
Maximum: 0.2132
Minimum: 0.2061

Table of predicted means for prov

prov	1	2	3	4	5	6	7	8
	2.422	3.143	2.865	2.260	3.719	3.545	3.947	2.747

prov	10	11	12	13	14	15	16	17
	2.646	2.050	2.750	2.983	2.653	3.446	3.921	3.429

prov	18	19	20	21	22	23	24	25
	3.362	3.654	3.369	3.221	3.332	3.245	3.693	3.846

prov	26	27	28	29	30	31	32	33
	3.523	3.527	3.100	3.900	3.545	4.129	4.276	3.743

prov	34	35	36	37	38	39	40	41
	3.889	4.118	3.853	3.454	3.038	3.531	3.741	3.742

prov	42	45	46	47	48	50	51	52
	3.876	4.213	3.466	4.469	3.750	2.557	3.457	3.323

prov	53	54	55	56	57	58	59	60
	3.593	3.534	2.716	2.770	3.981	2.997	3.398	3.012

prov	61	62	63
	3.224	3.281	3.575

Standard errors of differences

Average: 0.3483
Maximum: 0.3592
Minimum: 0.3327
Average variance of differences: 0.1213

Table 8.4 Continued

Table of predicted means for repl.row

row repl	1	2	3	4	5	6
1	3.532	4.167	4.077	3.642	3.578	3.530
2	3.172	3.370	3.170	3.108	3.059	2.882
3	3.047	3.248	2.922	3.122	3.471	3.466
4	3.683	3.589	3.536	3.437	3.471	3.298

Standard errors of differences or means cannot be formed using random terms when METHOD=sparse.

Table of predicted means for repl.column

column repl	1	2	3	4	5	6	7
1	3.761	3.992	4.180	4.173	4.164	4.089	3.771
2	3.454	3.277	3.588	3.543	3.105	3.722	3.132
3	3.330	3.205	3.697	3.676	3.597	3.448	3.007
4	3.476	3.595	3.938	3.921	3.513	3.873	3.313

column repl	8	9	10
1	3.920	3.384	2.180
2	2.966	2.915	1.569
3	3.432	3.056	1.678
4	3.745	3.356	2.293

There is no analysis of variance table available for the mixed-effects model because independent error assumptions underlying the analysis of variance table are no longer satisfied. In other words, we do not use the F-test to test for differences between the means for fixed-effects terms in the mixed-model analysis. Instead a quantity called the Wald statistic can be calculated; it is distributed as a X^2 statistic. For example, to test the significance of our 59 provenances, the Wald statistic is calculated as 284.06; the 0.1% point of the X^2 tables with 58 d.f. is 97 and so we can determine that provenance differences are highly significant. However, as mentioned in Section 8.2.3, we have not yet partitioned out the effects of countries; this is done in Section 8.4, below. An average standard error for the difference between two estimated provenance means can also be derived for fixed-effects terms. For comparison we have assembled in Table 8.5 the estimated provenance means and average standard errors for the difference between two provenance means for the RCB model and for the mixed-effects latinised row–column model. Note the reduction in the average standard error for the mixed-effects model; 63% of that for the RCB model.

Table 8.5 Table of estimated provenance means from analyses of the RCB and mixed-effects models for Example 8.1.

Country	Provenance	RCB	Mixed	Repl
Australia	1	1.85	2.42	4
Australia	2	3.24	3.14	4
Australia	3	2.94	2.86	4
Australia	4	2.50	2.26	4
Benin	5	3.34	3.72	4
China	6	3.38	3.54	4
China	7	3.98	3.95	4
China	63	3.70	3.58	4
Egypt	8	2.64	2.75	4
Egypt	10	2.47	2.65	4
Egypt	11	2.40	2.05	4
Fiji	12	2.39	2.75	4
Fiji	13	2.58	2.98	4
Fiji	56	2.87	2.77	4
Guam	14	2.34	2.65	4
India	15	3.77	3.45	4
India	16	3.60	3.92	4
India	17	3.45	3.43	4
India	18	3.26	3.36	4
India	19	3.83	3.65	4
India	20	3.55	3.37	4
Kenya	21	3.32	3.22	4
Kenya	22	3.55	3.33	4
Kenya	23	3.17	3.25	4
Kenya	24	3.64	3.69	4
Kenya	25	4.01	3.85	4
Kenya	26	3.74	3.52	4
Kenya	27	3.52	3.53	4
Kenya	28	2.99	3.10	4
Malaysia	29	4.34	3.90	4
Malaysia	30	3.82	3.54	4
Malaysia	31	4.39	4.13	4

Table 8.5 Continued

Malaysia	32	4.04	4.28	4
Malaysia	33	3.81	3.74	4
Malaysia	34	3.59	3.89	4
Malaysia	35	4.26	4.12	4
Malaysia	36	4.00	3.85	4
Malaysia	57	4.04	3.98	4
Mauritius	58	3.12	3.00	4
Philippines	38	3.46	3.04	4
Philippines	39	3.36	3.53	4
Philippines	40	4.02	3.74	4
PNG	37	3.65	3.45	4
Puerto Rico	59	3.35	3.40	4
Solomon Is.	41	3.49	3.74	4
Solomon Is.	42	3.91	3.88	4
Sri Lanka	60	3.37	3.01	4
Sri Lanka	61	3.02	3.22	4
Sri Lanka	62	3.35	3.28	4
Thailand	45	3.96	4.21	4
Thailand	46	3.20	3.47	4
Thailand	47	4.17	4.47	4
Thailand	48	4.04	3.75	4
Vanuatu	50	2.76	2.56	4
Vietnam	51	3.56	3.46	4
Vietnam	52	3.11	3.32	4
Vietnam	53	3.63	3.59	4
Vietnam	54	3.48	3.53	4
Vietnam	55	2.61	2.72	4
Average standard error		0.55	0.35	

The REML output in Table 8.4 includes values for rows and columns-within-replicates even though these are random effects, i.e. best linear unbiased predictions (BLUPs). In tree breeding, the use of mixed-effects models and BLUPs is well-established for prediction of genetic treatment effects (White *et al.* 2007, chapter 15). In Fig. 8.3 we plot the *repl.row* BLUPs from the Table 8.4 analysis. The first six values are the *repl.row* BLUPs from replicate 1, the next six from replicate 2 and so on. Because of the latinised design, it is appropriate to plot these BLUPs as a

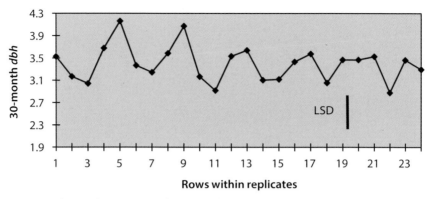

Fig. 8.3. Rows within replicates BLUPs for Example 8.1.

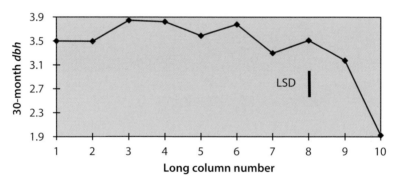

Fig. 8.4. Estimated long column means for Example 8.1.

sequence of 24 values. It is clear from Fig. 8.3 that there is considerable variation between rows within replicates and the mixed-model analysis has allowed us to separate this variation from the estimate of experimental error. The LSD (Section 4.2) at the 5% probability level has been included in Fig. 8.3 to provide a guide to the extent of the *repl.row* variation.

The estimated means for long columns can also be plotted to study the nature of the field trend (Fig. 8.4) and again the LSD at the 5% probability level is included. It is clear that growth in the middle columns of the trial is better than those on the outside. The growth in column 10 has been very poor, for an interesting reason. The trial was grown on land that had been mined for bauxite. The mining process in Weipa is such that an overburden of soil is removed to an already mined area and the thin layer of bauxite is scraped off. Hence there is a constant transfer of

Casuarina equisetifolia provenance trial at Weipa, Queensland (aged 12 months). The effect of waterlogging in the last long column (see Fig. 8.4) is evident on the left. (Photo Khongsak Pinyopusarerk)

overburden around the mine site. When the trial was laid out there was no obvious reason in the field for employing the latinised feature; it was done as a matter of course since the replicates were contiguous in the field and it did not cost us anything. After the trial was established, however, an extra overburden deposit area was created near to and parallel with column 10. A consequence was that in the wet season (the climate in northern Australia is monsoonal) there was an extensive build-up of water against the raised soil area and the trees in column 10 of the trial were partially submerged for some time. This clearly had an effect on their growth, but the long column blocking has allowed us to effectively adjust for this complication. The latinised feature has also ensured that no more than one plot of any of the seedlots was subjected to the effects of the water.

8.4 Nested treatment structure

In the majority of trials, the treatments are not simply an arbitrary set of entries to be compared. Rather, there is a structure or grouping. For example we might have a species/provenance structure or more commonly a provenance/progeny structure; these are examples of what is known as nested treatment structure. In Section 7.8

we mentioned incorporating nested treatment structure into experimental design by using the grouping facility in CycDesigN; here we will make some recommendations for analysis:

1. If we have v treatments grouped into a number of provenances say, it is usual to number the treatments from 1 to v and, as a secondary factor, record which group the treatment belongs to. Such a coding system can cause difficulties in the estimation of treatment means when we try to simultaneously accommodate blocking structures and the nested treatment structure. This is certainly the case in Genstat when there are incomplete blocks and we want to use either the **fit** (fixed-effects model) or **reml** (mixed-effects model) commands. Therefore, following a theme of this book, we suggest that an extra stage be introduced into the analysis. First, the analysis is carried out with the appropriate blocking structures, ignoring any nested structure. The estimated treatment means from this analysis and the numbers of replicates in which each treatment is present are then saved into a file for further analysis.

2. The estimated treatment means are then analysed to partition the treatment sum of squares according to the nested treatment structure that is present. The **anova** command in Genstat works very effectively for nested treatment structure, even with the usual case of unequal group sizes. If there are missing plots, a weighted analysis of variance can be carried out on the estimated treatment means using the treatment replicate numbers as weights (Appendix A). This is particularly important when there are a number of missing plots from a trial with many replicates – a typical scenario in single-tree plot experiments. When weights are used, there is no need to incorporate a scaling factor when trying to relate the nested treatment structure analysis to the main analysis at the *plots* level (from which we obtain the error term for testing treatment differences within groups).

3. Sometimes treatments are modelled as random effects in order to obtain variance components (Chapter 6) but are nested within a factor that is specified as fixed, e.g. Stackpole *et al.* (2010). Then it is more convenient to carry out the analysis in one hit, rather than the two-stage process discussed above.

Example 8.1 (continued)

The Genstat program in Table C8.3 prints out estimated provenance means for both the mixed model and RCB analyses; in Table 8.5 these have been enhanced with country codes given in Table 8.1. The estimated means can then be analysed according to this provenances-within-countries information. The Genstat program to analyse this nested provenance structure is given in Table C8.4. The resulting analysis of variance table is presented in Table 8.6; the residual mean square (0.195) from the analysis in Table 8.4 has been manually added to this table.

Table 8.6 Genstat output from the analysis of nested provenance structure for Example 8.1 with residual mean square manually added.

Analysis of variance

Variate: mixed
Weight variate: repl

Source of variation	d.f.	s.s.	m.s.	v.r.	F pr.
country	17	47.8213	2.813	14.43	<.001
country.prov	41	13.6472	0.333	1.71	0.013
Residual	119		0.195		

Tables of means

Variate: mixed
Weight variate: repl

Grand mean 3.40

country	Australia	Benin	China	Egypt	Fiji
	2.67	3.72	3.69	2.48	2.83
wt.rep.	16.00	4.00	12.00	12.00	12.00

country	Guam	India	Kenya	Malaysia	Mauritius
	2.65	3.53	3.44	3.94	3.00
wt.rep.	4.00	24.00	32.00	36.00	4.00

country	Philippines	PNG	Puerto Rico	Solomon Is.	Sri Lanka
	3.44	3.45	3.40	3.81	3.17
wt.rep.	12.00	4.00	4.00	8.00	12.00

country	Thailand	Vanuatu	Vietnam
	3.98	2.56	3.32
wt.rep.	16.00	4.00	20.00

Remember that we carried out a similar operation in Table 5.4 when adopting a multi-stage approach to analysis across sites.

We must, however, be aware that the resulting F-values for countries and provenances within countries are only approximate, having been derived from the REML mixed-effects model analysis. Nevertheless they provide a very good indication of the significance of the terms in the nested treatment structure. Given the statistical and computational difficulties associated with the estimation and testing in mixed-effects models with nested treatment structure, the above two-stage procedure and subsequent approximate F-tests provide the most convenient approach to this very common situation.

8.5 Treatments as random effects

Next, we examine a clonal trial which evaluates hybrid eucalypt clones produced by a breeding program. As with many genetic experiments, there are control treatments, against which we wish to benchmark the new clones. We also wish to estimate variance components so as to calculate clonal BLUPs and obtain an estimate of broad-sense heritability, but exclude the control treatments from this calculation.

Example 8.2

In Example 7.1 we discussed a *Eucalyptus* clone trial conducted in Vietnam and described the experimental layout. The trial tested 56 hybrid clones of the interspecific hybrid combination *E. urophylla* × *E. pellita* (UP). These candidates had been selected from progeny trials of control-pollinated hybrid families; here we ignore the parental origins of the different UP clones. In addition to the 56 UP clones, there were four control treatments: two commercially planted *Eucalyptus* clones named U6 and PN14, an imported hybrid *Eucalyptus* clone (UG323) and a bulk seedlot from an *E. urophylla* seed orchard. The trial aimed to rank the new clones and gauge their performance relative to the controls. The plot summary file (Section 3.2.3) for three-year diameter at breast height (*dbh*) is given in Table B8.2. A Genstat program for the analysis of *dbh* at the *plots* level is presented in Table C8.5.

The four control treatments (treatment numbers 1…4) have been specified as fixed effects, whereas the 56 clones for evaluation were designated as random effects. The data file and analysis code are structured to obtain the correct test for the comparison of controls versus clones. To do this an additional factor with two levels, contcompf, has been declared to enable comparison of two groups (the four controls and the 56 UP clones), creating a nested treatment structure. Furthermore contcompf must also be declared in variate form (contcompv) as explained in example 2 of Piepho *et al.* (2006).

The results of the analysis, including estimated means for fixed effects, comparisons between controls and the UP clones, variance components and BLUPs for the 56 UP clones, are presented in Table 8.7.

Table 8.7   Genstat output from the analysis of *dbh* means for Example 8.2.

REML variance components analysis

Response variate:	v[1]; dbh – diameter at breast height
Fixed model:	Constant + repl + column + contcompf + contcompf.standard
Random model:	repl.row + repl.column + contcompv.clone
Number of units:	300

Residual term has been added to model

Sparse algorithm with AI optimisation

Table 8.7   Continued

Covariates not centred

Estimated variance components

Random term	component	s.e.
repl.row	0.0802	0.0349
repl.column	0.0529	0.0327
contcompv.clone	0.4950	0.1126

Residual variance model

Term	Model(order)	Parameter	Estimate	s.e.
Residual	Identity	Sigma2	0.399	0.0437

Tests for fixed effects

Sequentially adding terms to fixed model

Fixed term	Wald statistic	n.d.f.	F statistic	d.d.f.	F pr.
repl	8.20	4	2.05	26.6	0.116
column	7.99	5	1.60	19.7	0.207
contcompf	72.66	1	72.66	254.0	<0.001
contcompf.standard	51.75	3	17.25	208.2	<0.001

Dropping individual terms from full fixed model

Fixed term	Wald statistic	n.d.f.	F statistic	d.d.f.	F pr.
repl	8.20	4	2.05	26.6	0.116
column	7.77	5	1.55	19.7	0.219
contcompf.standard	51.75	3	17.25	208.2	<0.001

Table of effects for contcompv.clone

clone	0	UP103	UP110	UP111	UP115
	0.0000	−0.4088	−1.6879	−0.9925	−0.3290
clone	UP122	UP123	UP124	UP127	UP135
	0.3754	0.3370	−0.2612	−0.9488	0.1725
clone	UP138	UP140	UP144	UP147	UP149
	0.0630	−0.2653	−0.4274	0.5385	0.6272
clone	UP150	UP151	UP153	UP154	UP156
	−1.1109	−0.0581	1.5114	−0.6518	−1.0035
clone	UP159	UP165	UP169	UP171	UP187
	−0.5466	0.0868	−0.4572	0.2862	−1.0069
clone	UP188	UP190	UP192	UP218	UP219
	0.3618	−0.2650	−0.2455	−0.1928	−0.2311
clone	UP220	UP223	UP225	UP227	UP233
	0.0225	0.3923	−0.2066	0.2154	−0.5030

clone	UP238	UP240	UP251	UP252	UP254
	−0.0018	−0.7820	0.6360	0.1748	0.8668

clone	UP262	UP266	UP274	UP276	UP278
	0.1608	0.3733	1.1723	1.0462	0.1415

clone	UP282	UP283	UP287	UP288	UP290
	0.8413	0.1317	0.4145	0.8466	−0.5332

clone	UP291	UP296	UP297	UP300	UP301
	0.4248	0.3845	0.2352	1.0195	0.6590

clone	UP316	UP317
	−0.6971	−0.7049

Table of predicted means for repl

repl	1	2	3	4	5
	7.662	7.957	8.029	8.146	8.232

Standard of error differences: 0.2167

Table of predicted means for column

column	1	2	3	4	5	6
	8.212	8.062	8.084	7.823	7.752	8.100

Standard errors of differences

Average: 0.2015
Maximum: 0.2026
Minimum: 0.2005

Table of predicted means for contcompf.standard

standard contcompf	0	PN14	SSOseed	U6	UG323
0	*	7.703	6.084	6.554	8.971
1	8.908	*	*	*	*

Standard errors of differences

Average: 0.3880
Maximum: 0.4362
Minimum: 0.3212

Average variance of differences: 0.1534

hmean	sigma2_t	sigma2_m	sigma2_c	h2
7.97771	1.76181	0.17839	0.49500	0.20327

The long columns are not significant although the estimated means hint at a drop in the middle columns. The inclusion of rows and columns within replicates, however, is worthwhile and reduces the error mean square from 0.553 (RCB analysis) to 0.399.

From the Wald tests comparing fixed effects means, the mean *dbh* of the 56 UP clones, 8.91 cm, is significantly (P<0.001) greater than the mean of the four controls, and the controls also differ significantly from one another. The *E. urophylla* pure-species seed orchard seedlot is the slowest growing of the four controls at 6.55 cm, while the single imported *E. urophylla* × *E. grandis* clone is the fastest at 8.97 cm. While the mean of the 56 UP clones has been estimated as a fixed effect, the *dbh* values of individual clones are predicted as random effects (BLUPs) centred around this mean.

From the variance components estimated by the model, we can calculate a broad sense 'clonal heritability' as follows:

$$\hat{H}^2 = \sigma_c^2 / \sigma_p^2 = 0.20 \tag{8.6}$$

Although the broad sense clonal heritability resembles the narrow-sense heritability calculated from progeny trials that was presented in (6.6), it is not used

Debarked acacia logs by the roadside near Hue, Central Vietnam. In the tropics, many smallholder plantations are harvested on short rotations, yielding small logs that are used for pulpwood.

directly to forecasting gain from deploying the best clones, and has no application to their merit for further breeding. It simply shows the ratio of genetic variance among the clones relative to the total phenotypic variance, thus providing a useful measure of the effectiveness of the test, and a simple way of comparing different tests; trials with a high broad-sense heritability show good potential for selecting superior clones (White *et al.* 2007, chapter 6). Note that the variance components for the rows and columns within replicates are not included in (8.6).

To calculate deployment gain from selecting the best of the 56 UP clones for deployment, a different parameter called the clonal mean repeatability can be estimated (Mullin and Park 1992; Kien *et al.* 2010). The application of clonal mean repeatability for predicting the deployment gain from selecting a sub-set of the best clones is not presented here. Instead, we show a simpler way to predict deployment gain by making use of the clone BLUPs generated by the analysis (see chapter 15 of White *et al.* 2007, for a detailed exposition of the application of BLUPs). The mean of the top 5 UP clone effects yields a predicted deployment gain of 1.12 cm, i.e. a 12.6% gain in *dbh* at 36 months relative to the average of the UP clones. We are also interested in the predicted deployment gain relative to the (fixed effect) mean of the two commercial clones, U6 and PN14, this is much greater, at 2.90 cm.

Of course, clone rankings and predicted gains from single-tree plots or line plots are not very reliable because of interspecific competition. It would be advisable to further test the best of the UP clones in second-stage clone proving trials against carefully chosen control treatments using large block plots of say 25, 36 or 49 trees, and examining not just *dbh* but an objective trait such as volume production at rotation age (Stanger *et al.* 2011).

8.6   Summary of analysis options

An objective of this book is to provide a framework for the analysis of field trials with multiple-tree plots, designed using appropriate blocking structures. A multi-stage process has been mapped out, the first stage being the production of a plot summary file (Section 3.2.3). Single-tree plot experiments are accommodated by making the *trees* level synonymous with the *plots* level, but in so doing, we have to forgo plot variance information which is an integral part of the plot summary file.

The analysis of variance is carried out at the *plots* level and depends on the type of experimental layout used, ranging from RCB analysis to the latinised row–column analysis employed in Example 8.1. Between these extremes there are many other possibilities, such as the other types of generalised lattice designs discussed in Chapter 7. Here we will discuss some common generalised lattice designs.

1. The simplest type of generalised lattice is where there is a single blocking structure within replicates (e.g. alpha designs) as described in Section 7.4. We normally include replicates in the model as fixed effects because:

    i.   Replicates often have a physical interpretation, e.g. extra treatments, such as levels of fertiliser, can be applied to replicates.

    ii. Replicates are orthogonal to treatments so, unless there are missing plots, there is no treatment information to be recovered by a random effects specification. Even with missing plots it is usually better to leave replicates as fixed effects, because the amount of treatment information that can be recovered is likely to be small.

    iii. From the computational point of view it is best to minimise the number of random effects. This is because the iterative process used to obtain REML estimates of variance components can lose accuracy and possibly fail to converge as the number of components increases.

2. If the plots in each replicate are in a rectangular arrangement, the experiment should always be designed for possible row–column adjustment. This applies regardless of whether we have square, rectangular or one-dimensional configurations of trees within plots. If after estimating row- and column-within-replicate variance components, one of them is close to zero, the relevant blocking factor can be excluded from the model and the analysis redone. Very little, if anything has been lost by initially including a blocking factor that is ultimately not needed, especially when it is not always easy to decide at the design stage the nature and direction of any field trend. A package such as CycDesigN makes it possible to incorporate as many blocking structures as appropriate for the particular field layout.

   On the other hand, even if trials have not been designed with incomplete blocking structures, a mixed-model analysis can be performed if it is appropriate for the field layout. We just have to accept the possibility of poor design leading to loss of efficiency (and hence precision). But if field trend within replicates is strong, the use of post-blocking can still be very worthwhile (Patterson and Hunter 1983). Further adjustment for within-replicate field trends may be appropriate, using spatial analysis (Piepho et al. 2022).

3. Latinised designs (Section 7.5.1) form an important and very common design arrangement. It is usually valuable to consider row and column blocking structures within replicates, together with the long columns running down the replicates as discussed in Section 7.5.3 and demonstrated in Example 8.1.

   The long columns are usually specified as fixed effects for similar reasons to those given earlier for replicates, namely:

i.   If the majority of the treatments are represented in long columns there is not much treatment information to be recovered from comparisons between them.

ii. Long columns can have a physical interpretation such, as watering channels in cotton trials (Williams 1986).

The one-dimensional replicate structure (i.e. contiguous replicates in a line) in Fig. 7.6 is usually preferable to a two-dimensional replicate structure, e.g. Fig. 7.7. This is because having long columns running down all the replicates allows the seedlots to be made more nearly orthogonal to the long columns. For designs with many contiguous replicates, however, a two-dimensional arrangement of replicates can be sensible. For example, as part of an ACIAR project in Thailand, provenance/progeny trials for 315 *Eucalyptus camaldulensis* seedlots were established at three sites in 1991. As it was planned for the trials to be later converted to seedling seed orchards, 15 replicates of three-tree plots were used. The replicates, each consisting of 21 rows and 15 columns, were laid out in a 3 × 5 arrangement.

A difficulty with the two-dimensional arrangement of contiguous replicates is that a number of possible blocking structures are created from the partition of the replicate factor into components for column and row replicates. Hence it is necessary to assess the importance of a number of variance components and exclude superfluous terms from the final specification of the mixed-effects model.

Appendix A: Introduction to Genstat

A.1 Introduction

This book uses the statistical package Genstat (VSN International) for data analysis. Other good options include SAS and R.

Genstat is a powerful package which allows users to perform intricate data management operations and detailed statistical analyses. Casual users, however, may have insufficient background knowledge to employ the package effectively. To reduce this problem, we have given specific rules for the construction of data files (Section 3.2.4) so that the need for data manipulation is minimised.

In this Appendix we give a selection of the Genstat commands and symbols used in the book. These are only a very small subset of the commands available but they demonstrate how much can be achieved by following the above strategy. Our list excludes some extra commands used in Tables C5.3 and C5.4 for the analysis of the complete and incomplete site-by-seedlot tables respectively. The Genstat programs for these tables have been set up like subroutines (in Genstat terminology, procedures) and we would not recommend that anything but the specification of the size of the table, site and treatment names, and input details be changed.

The Genstat commands discussed in the next section have the general form:

command [options] parameters

where there can be a number of options and/or parameters for each command. Our examples will cover only options and parameters actually used in the book, although in some cases not even all options and parameters given here are used.

When options or parameters are not specified, default values are used; these are listed in detail in the package documentation.

A.2   Genstat commands

A.2.1   Special symbols

"   Double quotes are used to enclose comments within the Genstat program.

:   A colon signifies the end of a data set, either in the Genstat program or in an input file.

&   An ampersand repeats the previous statement name.

\   A backslash is the continuation symbol to indicate that the current statement continues into the next line.

A.2.2   Set-up

units [nvalues = 16]

   Specifies the default length of structures such as data vectors. In this case there would be 16 items in the data file.

text [values = SO,P] seedname

   A convenient way to specify the actual names of the levels of a factor. Sometimes these may be seedlot numbers, in which case the numbers have to be put in single quotes, e.g. '13877'.

scalar plotmn

   A single value, such as the mean of the trees in a plot.

variate pc3; extra = !t('percentage germination 15/8/92')

   Declares a variate called *pc3*. If the number of values is not specified, the default given by the **units** command is assumed. The **extra** parameter allows extra text to be associated with the variate. Here the full variate would read

   pc3 percentage germination 15/08/92

   Another useful form is

variate [values = 1...5] lev[1]

   Defines the variate *lev[1]* with values from 1 to 5.

factor [levels = 2; labels = seedname] seedlot

   Declares a factor called *seedlot* which has two levels. The actual names of the levels are given by the text variate, i.e. SO and P.

   Another useful form is

factor [levels = lev[1]] seedlot; decimals = 0

   Defines the *seedlot* levels according to the values of the variate *lev[1]*, i.e. the values from 1 to 5; on the output the levels would be printed with no decimal places.

stop

   The end of the Genstat program.

A.2.3 Input/output commands

import 'ex3.1a.xlsx'

> Enters experimental structures (e.g. factors and data files) from Excel into Genstat. The files should be in the recommended form, i.e. column by column for each structure and be headed with the structure name.

groups [redefine=yes] repl, plot, tree, seedlot

> Coverts structures read from Excel into Genstat factors.

open 'ex3_1sum.dat'; channel = 3; filetype = output; width=200

> Identifies the file *ex3_1sum.dat* with output channel number 3 in the Genstat program; the line length can be up to 200 characters.

read [channel = 2; end = *] repl

> Reads the structure *repl* from the file identified with input channel number 2. If the **units** command has been used to specify the length of *repl*, then the **end** = * option avoids us having to terminate the data file with the : symbol.

print [channel = 3; iprint = *] plotmn; fieldwidth = 12; decimals = 5

> Prints the structure *plotmn* to the file identified with output channel number 3. using a fieldwidth of 12 spaces with 5 decimal places. The title for *plotmn* is suppressed using **iprint** = *.

close channel = 3; filetype = output

> Closes off the channel number 3 which was opened for output.

A.2.4 Manipulation

for i = v[1...3]

> Introduces a loop; the statements within the loop are executed for i = v[1], then for i = v[2] and finally for i = v[3].

endfor

> The end of the loop.

calculate plotmn = mean (ht)

& plotvar = var (ht)

> Calculates the mean and variance of the elements of the variate *ht*.

calculate plotcnt = sum (index)

> Calculates the sum of the elements of the variate *index*.

calculate contcomp = newlevels (treat; !(1(1),3(2)))

> If the factor *contcomp* is declared with 2 levels, the **newlevels** function identifies the first level of *contcomp* with the first level of *treat* and the second level of *contcomp* with levels 2, 3 and 4 of *treat*.

tabulate [classification = seedlot] v[1]; nobservations = trepl

> Produces a table of seedlot replication numbers for the variate *v[1]*, i.e. the table *trepl* contains the number of replicates for which each seedlot is present.

restrict v[1]; condition = treat.ni.!(1,2)
   Excludes treatment levels 1 and 2 from any operations on the data in the
   variate *v[1]*.

A.2.5 Orthogonal analysis of variance

block repl / plot
   Defines a blocking structure with two strata, the *repl* stratum and the *repl.plot*
   stratum. The notation *repl / plot* is known as a nested structure and can be
   expanded into *repl + repl.plot*.
treatment irrig * fert
   Defines a treatment structure for the factors *irrig, fert* and their interaction
   irrig.fert. The notation *irrig * fert* is known as a crossed structure and can be
   expanded into *irrig + fert + irrig.fert*.
anova [fprobability = yes; weight = trepl] ht; fittedvalues = fitted; residuals = resid
   Performs a weighted analysis of variance of the variate *ht* according to the
   specified block and treatment declarations, using the variate *trepl* as weights.
   The analysis of variance table includes F-probabilities for each treatment term
   that can be tested against an appropriate residual mean square. The fitted
   values and residuals from the analysis are saved in the variates *fitted* and *resid*
   respectively. The treatment mean squares in **anova** are only approximate when
   there are missing values; exact values are given by using the **fit** command.
aplot fitted
   After the above anova command a scatterplot is produced with *resid* on the
   y-axis and *fitted* on the *x*-axis.

A.2.6 Non-orthogonal analysis of variance

model ht
   Defines the response variate *ht* for analysis of variance using a regression
   approach.
fit [print = model, accumulated; fprobability = yes] site + treatment
   Fits the linear regression model, fitting first the factor *site* then the factor
   treatment. The model specification is printed out and a summary analysis of
   variance table is produced. For orthogonal designs this table is the same as that
   given by **anova**.

A.2.7 Mixed-effects model analysis

vcomponents [fixed = repl; absorb = repl] random = treatment;
   constraints=positive
   Defines a model for residual maximum likelihood (REML) analysis with the
   factor *repl* as fixed effects and the factors *treatment* as random effects. An

absorbing factor (*repl*) is used to save computational time and space. The estimated variance component for *treatment* is prevented from being negative.

reml [print = model, components, means, waldtests; pterms = repl + treatment] v[1]
Fits the REML model for the variate *v[1]* according to the specification in **vcomponents**. The model, variance components, estimated means for *repl* and best linear unbiased predictions (BLUPs) for *treatment* are printed out along with the Wald statistic for testing *repl*.

vkeep [sigma2 = s2] treatment; means = tmean[1]; components = sigma2_f
Copies the residual mean square and treatment variance component from a REML analysis into the scalars *s2* and *sigma2_f* respectively. These can then be used for further calculations in the Genstat program. The estimated means from the analysis are stored in the table *tmean[1]*.

Glossary

Additive genetic variance – the part of genetic variance due to additive effects of genes, as opposed to non-additive effects such as dominance and epistasis (see Falconer and Mackay 1996).

Age–age correlation – the correlation between a trait measured at one age with the same trait measured at another age. It may be a **genetic correlation**.

Alpha array – an $r \times k$ array of numbers used to develop an **alpha design** for r replicates of sk treatments (see John and Williams 1995).

Alpha designs – a class of **generalised lattice designs** generated from an **alpha array**.

Best linear unbiased estimates (BLUEs) – values for terms specified as **fixed effects** in a **linear model**.

Best linear unbiased predictions (BLUPs) – values for terms specified as **random effects** in a **mixed-effects model**.

Blocking structures – physical (perhaps artificial) groupings of experimental units related to the **experimental design**.

Bulk seedlots – seedlots made by mixing together seedlots from different mother trees.

Clone – a set of trees (**ramets**) propagated from a founder tree (ortet) without going through the sexual process, i.e. by vegetative means or by micropropagation. The ortet and the ramets comprise the clone, and are genetically identical.

Coefficient of relationship (r) – within a family, this is the degree of relationship among members of the family; for half-sibs it is 1/4 and for full-sibs, 1/2.

Combined selection index – a selection index is a (usually linear) combination of measured values to make up a single index value for each individual in a set. A combined index also combines information from relatives.

Completely randomised designs – designs with no **blocking structures**.

Concurrence matrix – a matrix describing the number of occurrences of pairs of treatments within the same **blocking structure**.

Contiguous replicates – where **replicates** of an experiment are adjacent, sometimes resulting in a two-dimensional array of plots over the entire experimental area.

Control-pollinated families – families produced by controlled pollination, sometimes (but not always) in some special pattern.

Degrees of freedom (d.f.) – corresponds to the number of independent **parameters** estimated for factors in a **fixed-effects model**.

Design matrix – a matrix of zeros and ones in which columns represent factor levels and rows represent experimental units.

Efficiency factor (E) – the ratio of the average pairwise variance between treatments for a design compared with the pairwise variance for a **randomised complete block** design.

Elimination trial – usually a short-term trial to eliminate species, varieties, seedlots or clones from further consideration.

Experimental design – the layout of an experiment. This includes the choice of **blocking structures**, taking into account the site characteristics and the available treatment and replication numbers.

External replication – usually the number of replicates, as in **randomised complete block designs** (c.f. **internal replication**).

Factor (factor levels) – a Genstat **blocking** or **treatment structure**. Each factor can have several levels.

Factorial designs – experimental designs with more than one **treatment structure** applied to each plot in such a way that interactions can be estimated.

Field trend – environmental variation between one part of an experimental site and another.

Fitted values – values for each experimental unit derived from combining estimated effects from the model.

Fixed effects – model terms where a parameter is specified for each factor level. Inferences are made only about the experimental material itself and not about any hypothetical population of which the experimental material is a sample (c.f. **random effects**).

Fixed-effects model – a **linear model** in which there are only **fixed effects** and residual terms.

Generalised lattice designs – a class of resolvable designs with one or more incomplete **blocking structures** (see **square** and **rectangular lattice designs**).

Genetic correlation – an expression of the amount of genetic change in one trait made when selecting for another.

Genotype – the genes carried by an individual. It is sometimes used to describe groups of individuals with a degree of genetic similarity. This is correct for clones, all individuals of which have the same genotype, but not strictly so for families within which there is much less genetic similarity.

Genotype-by-environment interaction – the differential response by genotypes to the environments in which they are grown. Seedlot-by-site interaction is a specific case.

Half-sibs – siblings (brothers or sisters) who have one parent in common (see **open-pollinated families**).

Heritability – (narrow sense) the proportion of the **phenotypic variance** accounted for by **additive genetic** differences among individuals.

Incidence matrix – a matrix where the columns and rows represent the levels of block and treatment factors respectively. The elements of the incidence matrix record the number of times treatments appear in blocks.

Incomplete block designs – designs with a **blocking structure** where the number of plots per block is less than the total number of treatments.

Internal replication – usually describes the number of trees per plot in forestry experiments (c.f. **external replication**).

Joint regression analysis – a form of analysis in which interactions are modelled by **linear regression** of individual treatment means on site means.

Latinised designs – resolvable incomplete block designs or, more commonly, resolvable row–column designs where the **replicates** are **contiguous**.

Least significant difference (LSD) – the minimum difference between two estimated means in order to attain significance at a prescribed probability level.

Linear model – a model to approximate data as the sum of a number of **parameters**.

Linear regression – a **linear model** of the form $y = a + bx$ fitted to experimental data. The a is a constant and the b is the **regression coefficient** relating an independent **variate**(x) with the dependent **variate** (y).

Long column – where **replicates** are **contiguous**, long columns are formed by combining columns from adjacent **replicates**.

Mixed-effects model – one in which there are both **fixed effects** and **random effects**.

Multiple-tree plot – an experimental plot consisting of more than one tree, either as a line plot or a two-dimensional arrangement.

NB&ED design – a spatial design enhanced with neighbour balance and evenness of distribution of treatments.

Non-contiguous plots – with **multiple-tree plots**, trees of the same seedlot are normally planted in adjacent positions. Non-contiguous plots occur when the trees which form the plot are not planted together, but are scattered at random throughout a **replicate**.

Open-pollinated families – these are formed when seed is collected from individual mother trees following uncontrolled pollination.

Orthogonal designs – designs are called orthogonal if all elements of the **incidence matrix** are the same, i.e. when all the treatments appear together the same number of times in each block.

Pairwise variance – the variance of the difference between two estimated means.

Parameters – terms in a mathematical model which can be estimated from the observed data.

Partially replicated designs – designs where a percentage of the entries for evaluation are replicated at a site whilst the remainder are unreplicated.

Phenotype – the physical appearance of an individual.

Phenotypic variance – the total variance ascribable to both genetic and environmental sources. It is usually calculated as the sum of the family, between-plot and within-plot variance but this depends on the experimental design, e.g. it would include interaction between fertilisers and seedlots in a fertiliser trial.

Plot variance – the variance between trees within a plot.

Provenance/progeny trials – provenance trials in which mother-tree identity of seedlings is retained.

Ramets – members of a **clone**.

Random effects – model terms where only a single **parameter** is estimated describing the variation between **factor** levels (c.f. **fixed effects**).

Randomised complete block (RCB) designs – **orthogonal designs** where treatments are randomised to the plots in each **replicate**.

Rectangular lattice designs – a special type of **generalised lattice design** for $s(s-1)$ treatments, where s is an integer.

Regression coefficient – a **parameter** appearing as a multiplier for an independent **variate** or another set of **parameters** in a mathematical model.

Replicate – a group of plots consisting of a single plot of each of the treatments (c.f. **external replication**).

Residual maximum likelihood (REML) estimation – a method of **parameter** estimation where variance components are estimated before **fixed effects**.

Residual mean square – a term in the analysis of variance table corresponding to the variance of the residuals.

Residuals – the difference between **fitted values** and observed data points in a **fixed-effects model**.

Resolvable designs – **experimental designs** where the plots can be grouped into discrete **replicates**.

Row–column designs – experimental designs with **blocking structures** (rows and columns) at right angles.

Seedlot – in this book we have often used the term 'seedlot' to refer to the genetic entities under test. Each seedlot may represent a whole provenance, or a single **open-** or **control-pollinated family** so equally, we could have used the terms 'clones', 'families', 'provenances' or 'species'. Also we have sometimes used the term 'seedling' to refer to young trees under test, without wishing to exclude **ramets** of a **clone**.

Selection index – see **Combined selection index**.

Selection intensity (i) – the difference between the mean of a selected population and the overall mean, divided by the phenotypic standard deviation.

Selfing – the process of self-pollination.

Single-tree plot – an experimental plot consisting of just one tree.

Spatial designs – designs where the spatial separation of treatments (within blocks or rows and columns) has been taken into account.

Species/provenance trials – species trials in which provenance identity is retained as well as species identity.

Split-plot designs – **factorial designs** where **treatment structures** are tested at different **strata**.

Square lattice designs – a special type of **generalised lattice design** for s^2 treatments, where s is an integer.

Standard error of a difference – the standard error for the comparison of two estimated treatment means.

Standard treatment – A treatment introduced into a partially replicated design mainly to provide a reference point for new entries.

Strata – levels of variation in an experimental design determined by the component **blocking structures**. Different strata are randomised separately.

Transformation – a function of a **variate** (e.g. square root, logarithm) usually taken to help satisfy mathematical model assumptions, such as equal variances of observations.

Treatment structures – structures to specify experimental quantities to be compared, such as different seedlots, levels of fertiliser application etc.

Variance component – a **parameter** to measure the variation between **factor** levels for **random effects**.

Variance ratio (F-test) – the ratio of a test mean square to a residual mean square. If the ratio is greater than tabulated critical values (usually at the 5% probability level) of the F-distribution, the test mean square is significantly greater than the residual mean square (at the 5% probability level).

Variate – a term used for a vector (or set) of data values.

Wald statistic – an approximate X^2 statistic for testing fixed effects in a mixed-effects model analysis.

Weighted analysis – an analysis of variance of a **variate** whose elements are assigned different levels of importance.

References

Apiolaza LA, Gilmour AR, Garrick DJ (2000) Variance modelling of longitudinal height data from a *Pinus radiata* progeny test. *Canadian Journal of Forest Research* **30**, 645–654. doi:10.1139/x99-246

Barnes RD, Matheson AC, Mullin LJ, Birks J (1987) Dominance in a metric trait of *Pinus patula* Scheide and Deppe. *Forest Science* **33**, 809–815.

Becker WA (1992) *Manual of Quantitative Genetics.* Academic Enterprises, Pullman.

Boland DJ, Brooker MIH, Turnbull JW, Kleinig DA (1980) *Eucalyptus Seed.* CSIRO, Melbourne.

Booth TH, Jones PG (1998) Identifying climatically suitable areas for growing particular trees in Latin America. *Forest Ecology and Management* **108**, 167–173. doi:10.1016/S0378-1127(98)00223-0

Brown AHD, Matheson AC, Eldridge KG (1975) Estimation of the mating system of *Eucalyptus obliqua* L'Hérit by using allosyme polymorphisms. *Australian Journal of Botany* **23**, 931–949. doi:10.1071/BT9750931

Burdon RD (1977) Genetic correlation as a concept for studying genotype–environment interaction in forest tree breeding. *Silvae Genetica* **26**, 168–175.

Burley J, Wood PJ (1976) A manual on species and provenance research with particular reference to the tropics. Tropical Forestry Papers No. 10. Commonwealth Forestry Institute, Oxford.

Bush DJ, Kain D, Matheson AC, Kanowski P (2011) Marker-based adjustment of the additive relationship matrix for estimation of genetic parameters – an example using *Eucalyptus cladocalyx. Tree Genetics and Genomes* **7**, 23–35. doi:10.1007/s11295-010-0312-z

Bush DJ, Kain D, Kanowski P, Matheson AC (2015) Genetic parameter estimates informed by a marker-based pedigree: a case study with *Eucalyptus cladocalyx* in southern Australia. *Tree Genetics and Genomes* **11**, 798–813. doi:10.1007/s11295-014-0798-x

Butcher PA, Glaubitz JC, Moran GF (1999) Applications for microsatellite markers in the domestication and conservation of forest trees. *Forest Genetic Resources Information* **27**, 34–42.

Cochran WG, Cox GM (1957) *Experimental Designs*, 2nd edn. Wiley, New York.

Cornelius J (1994) Heritabilities and coefficients of additive genetic variation in forest trees. *Canadian Journal of Forest Research* **24**, 372–379. doi:10.1139/x94-050

Cullis BR, Smith A, Coombes N (2006) On the design of early generation variety trials with correlated data. *Journal of Agricultural, Biological, and Environmental Statistics* **11**, 381–393. doi:10.1198/108571106X154443

Dickerson GE (1969) Techniques for research in quantitative animal genetics. In *Techniques and Procedures in Animal Science Research*, pp. 36–79. American Society of Animal Science, Albany, New York.

Dieters MJ, White TL, Littell RC, Hodge GR (1995) Application of approximate variances of variance components and their ratios in genetic tests. *Theoretical and Applied Genetics* **91**, 15–24. doi:10.1007/BF00220853

Digby PGN (1979) Modified joint regression analysis for incomplete variety x environment data. *The Journal of Agricultural Science* **93**, 81–86. doi:10.1017/S0021859600086159

Eldridge KG, Davidson J, Harwood CE, van Wyk G (1993) *Eucalypt Domestication and Breeding*. Clarendon, Oxford.

Falconer DS, Mackay TFC (1996) *Introduction to Quantitative Genetics*, 4th edn. Longman, Harlow.

Federer WT (1956) Augmented (or hoonuiaku) designs. *Hawaiian Planters' Record* **LV(2)**, 191–208.

Finlay KW, Wilkinson GN (1963) The analysis of adaption in a plant-breeding programme. *Australian Journal of Agricultural Research* **14**, 742–754. doi:10.1071/AR9630742

FWPA (2022) Australian tree breeding research revolutionising the industry. Forest & Wood Products Australia, Melbourne. <https://fwpa.com.au/australian-tree-breeding-research-revolutionising-the-industry/>.

Green JW (1971) Variation in *Eucalyptus obliqua* L'Hérit. *New Phytologist* **70**, 897–909. doi:10.1111/j.1469-8137.1971.tb02590.x

Griffin AR, Cotterill PP (1988) Genetic variation in growth of outcrossed, selfed and open-pollinated progenies of *Eucalyptus regnans* and some implications for breeding strategy. *Silvae Genetica* **37**, 124–131.

Harwood CE, Williams ER (1992) A review of provenance variation in the growth of *Acacia mangium*. In *Breeding Technologies for Tropical Acacias*. (Eds LT Carron, K Aken) pp. 22–30. ACIAR Proceedings No. 37. Australian Centre for International Agricultural Research, Canberra.

Hodge GR, Volker PW, Potts BM, Owen JV (1996) A comparison of genetic information from open-pollinated and control-pollinated progeny tests in two eucalypt species. *Theoretical and Applied Genetics* **92**, 53–63. doi:10.1007/BF00222951

John JA, Williams ER (1995) *Cyclic and Computer Generated Designs*, 2nd edn. Chapman & Hall, London.

John JA, Williams ER (1998) t-Latinized designs. *Australian and New Zealand Journal of Statistics* **40**, 111–118.

Kempthorne O, Nordskog AW (1959) Restricted selection indices. *Biometrics* **15**, 10–19.

Kendall MG, Stuart A, Ord JK (1987) *Kendall's Advanced Theory of Statistics, Vol 3: Design and Analysis, and Time Series.* Oxford University Press, New York.

Kien ND, Jansson G, Harwood CE, Almqvist C (2010) Clonal variation and genotype by environment interactions in growth and wood density in *Eucalyptus camaldulensis* at three contrasting sites in Vietnam. *Silvae Genetica* **59**, 17–28. doi:10.1515/sg-2010-0003

Lambeth CC (1980) Juvenile–mature correlations in *Pinaceae* and implications for early selection. *Forest Science* **26**, 571–580. doi:10.1093/forestscience/26.4.571

Lerner IM (1958) *The Genetic Basis of Selection.* Wiley, New York.

Libby WJ, Cockerham CC (1980) Random non-contiguous plots in interlocking field layouts. *Silvae Genetica* **29**, 183–190.

Macqueen DJ (1993) Exploration and collection of *Calliandra calothyrsus*. Final report of ODA project R4585. Oxford Forestry Institute, Oxford.

Matheson AC, Mullin L (1986) Variation among neighbouring and distant provenances of *Eucalyptus grandis* & *E. tereticornis* in Zimbabwean field trials. *Australian Forest Research* **17**, 233–250.

Matheson AC, Bell JC, Barnes RD (1989) Breeding systems and genetic structure in some Central American pine populations. *Silvae Genetica* **38**, 107–113.

McDonald MW, Brooker MIH, Butcher PA (2009) A taxonomic revision of *Eucalyptus camaldulensis* (Myrtaceae). *Australian Systematic Botany* **22**, 257–285. doi:10.1071/SB09005

McDonald MW, Maslin BR (2000) Taxonomic revision of the salwoods: *Acacia aulacocarpa* Cunn. ex Benth. and its allies (Leguminosae: Mimosoideae: section Juliflorae). *Australian Systematic Botany* **13**, 21–78.

Möhring J, Williams ER, Piepho HP (2014) Efficiency of augmented p-rep designs in multi-environmental trials. *Theoretical and Applied Genetics* **127**, 1049–1060. doi:10.1007/s00122-014-2278-y

Moran GF, Forrester RI, Rout AF (1990) Early growth of *Eucalyptus delegatensis* provenances in four field trials in south-eastern Australia. *New Zealand Journal of Forestry Science* **20**, 148–161.

Moran GF, Muona P, Bell JC (1989) Breeding systems and genetic diversity in *Acacia auriculiformis* and *A. crassicarpa*. *Biotropica* **26**, 250–256. doi:10.2307/2388652

Mullin TJ, Park YS (1992) Estimating genetic gains from alternative breeding strategies for clonal forestry. *Canadian Journal of Forest Research* **22**, 14–23. doi:10.1139/x92-003

Namkoong G, Kang HC, Brouard JS (1988) *Tree Breeding: Principles and Strategies.* Springer-Verlag, New York.

Ng MP, Williams ER (2001) Joint-regression analysis for incomplete two-way tables. *Australian and New Zealand Journal of Statistics* **43**, 901–906.

Otegbye GO, Samarawira I (1992) Genetics of growth and quality characteristics of *Eucalyptus camaldulensis* Dehnh. *Silvae Genetica* **41**, 249–252.

Palmberg C (1981) A vital fuelwood gene pool is in danger. *Unasylva* **33**, 22–30.

Patterson HD, Hunter EA (1983) The efficiency of incomplete block designs in National List and Recommended List cereal variety trials. *The Journal of Agricultural Science* **101**, 427–433. doi:10.1017/S002185960003776X

Patterson HD, Silvey V (1980) Statutory and recommended list trials of crop varieties in the United Kingdom (with discussion). *Journal of the Royal Statistical Society* **A143**, 219–252. doi:10.2307/2982128

Patterson HD, Thompson R (1971) Recovery of interblock information when block sizes are unequal. *Biometrika* **58**, 545–554. doi:10.2307/2334389

Patterson HD, Williams ER (1976) A new class of resolvable incomplete block designs. *Biometrika* **63**, 83–92. doi:10.2307/2335087

Piepho HP, Boer MP, Williams ER (2022) Two-dimensional P-spline smoothing for spatial analysis of plant breeding trials. *Biometrical Journal* **64**, 835–857. doi:10.1002/bimj.202100212

Piepho HP, Williams ER, Fleck M (2006) A note on the analysis of designed experiments with complex treatment structure. *HortScience* **41**, 446–452. doi:10.21273/HORTSCI.41.2.446

Piepho HP, Williams ER, Michel V (2021) Generating row-column field experimental designs with good neighbour balance and even distribution of treatment replications. *Journal of Agronomy and Crop Science* **207**, 745–753. doi:10.1111/jac.12463

Pinyopusarerk K, Doran JC, Williams ER, Wasuwanich P (1996) Variation in growth of *Eucalyptus camaldulensis* provenances in Thailand. *Forest Ecology and Management* **87**, 63–73. doi:10.1016/S0378-1127(96)03835-2

Pinyopusarerk K, Williams ER (2000) Range-wide provenance variation in growth and morphological characteristics of *Casuarina equisetifolia* grown in Northern Australia. *Forest Ecology and Management* **134**, 219–232. doi:10.1016/S0378-1127(99)00260-1

Raymond CA, MacDonald AC (1998) Where to shoot your pilodyn: within-tree variation in basic density in plantation-grown *Eucalyptus globulus* and *E. nitens* in Tasmania. *New Forests* **15**, 205–221. doi:10.1023/A:1006544918632

Rezende GDSP, de Resende MDV, de Assis TF (2014) *Eucalyptus* Breeding for Clonal Forestry. In *Challenges and Opportunities for the World's Forests in the 21st Century. Forestry Sciences.* (Ed. T Fenning) p. 81. Springer, Dordrecht.

Robinson GK (1991) That BLUP is a good thing: the estimation of random effects (with discussion). *Statistical Science* **6**, 15–51. doi:10.1214/ss/1177011926

Scheffé H (1959) *The Analysis of Variance.* Wiley, New York.

Sedgley M, Griffin AR (1989) *Sexual Reproduction of Tree Crops.* Academic Press, London.

Shelbourne CJA (1992) Genetic gains from different kinds of breeding and propagation populations. *South African Forestry Journal* **160**, 49–65.

Snedecor GW, Cochran WG (1989) *Statistical Methods*, 8th edn. Iowa State University, Ames.

Son DH, Harwood CE, Kien ND, Griffin AR, Thinh HH, Son L (2018) Evaluating approaches for developing elite *Acacia* hybrid clones in Vietnam: towards an updated strategy. *Journal of Tropical Forest Science* **30**, 476–487.

Speed TP, Williams ER, Patterson HD (1985) A note on the analysis of resolvable block designs. *Journal of the Royal Statistical Society* **B47**, 357–361.

Squillace AE (1974) Average genetic correlations among offspring from open-pollinated forest trees. *Silvae Genetica* **23**, 149–156.

Stackpole DJ, Vaillancourt RE, Downes GM, Harwood CE, Potts BM (2010) Genetic control of kraft pulp yield in *Eucalyptus globulus*. *Canadian Journal of Forest Research* **40**, 917–927. doi:10.1139/X10-035

Stanger TK, Galloway GM, Retief ECL (2011) Final results from a trial to test the effect of plot size on *Eucalyptus* hybrid clonal ranking in coastal Zululand, South Africa. *Southern Forests* **73**, 131–135. doi:10.2989/20702620.2011.639492

Stewart JL, Allison GE, Simons AJ (Eds) (1996) *Gliricidia sepium*: Genetic resources for farmers. *Tropical Forestry Papers* No. 33, Oxford Forestry Institute, Oxford.

Turnbull JW, Griffin AR (1986) The concept of provenance and its relationship to infraspecific classification in forest trees. In *Infraspecific Classification of Wild and Cultivated Plants*. (Ed. BT Styles) pp. 157–189. Clarendon, Oxford.

White TL, Adams WT, Neale DB (2007) *Forest Genetics*. CABI International, Massachusetts.

Willan RL, Hughes CE, Lauridsen EB (1990) Seed collection for tree improvement. In *Tree Improvement of Multipurpose Species*. (Eds N Glover, N Adams) pp. 11–37. Winrock International, Arlington, Bangkok.

Williams ER (1986) Row and column designs with contiguous replicates. *Australian Journal of Statistics* **28**, 154–163.

Williams ER, John JA, Whitaker D (1999) Example of block designs for plant and tree breeding trials. *Australian and New Zealand Journal of Statistics* **41**, 277–284. doi:10.1111/1467-842X.00082

Williams ER, Luangviriyasaeng V (1989) Statistical analysis of tree species trials and seedlot:site interaction in Thailand. In *Trees for the Tropics*. (Ed. DJ Boland) Chapter 14, pp. 145–152. ACIAR, Canberra.

Williams ER, Luckett DJ, Reid PE, Thomson NJ (1992) Comparison of locations used in cotton-breeding trials. *Australian Journal of Experimental Agriculture* **32**, 739–746. doi:10.1071/EA9920739

Williams ER, Piepho, HP (2019) Error variance bias in neighbour balance and evenness of distribution designs. *Australian and New Zealand Journal of Statistics* **61**, 466–473. doi:10.1111/anzs.12277

Wright JW (1976) *Introduction to Forest Genetics*. Academic Press, New York.

Yates F, Cochran WG (1938) The analysis of groups of experiments. *The Journal of Agricultural Science* **28**, 556–580. doi:10.1017/S0021859600050978

Zobel BJ, Talbert J (1984) *Applied Forest Tree Improvement*. Wiley, New York.

Index